Core Statistics

Based on two courses for new graduate students, Core Statistics provides concise coverage of the fundamentals of inference for parametric statistical models, including both theory and practical numerical computation. The book considers both frequentist maximum likelihood estimation and Bayesian stochastic simulation, focusing on general methods applicable to a wide range of models and emphasizing the common questions addressed by the two approaches.

This compact book aims to cover the core knowledge needed by beginning graduate students in statistical subjects, at a level suitable for those going on to develop new methods or undertaking novel applications of statistical modelling: Bayesian and frequentist approaches to modelling and inference; practical computational implementation, including numerical issues; brief coverage of some essential probability; and the essentials of R as a statistical programming language. Aimed also at any quantitative scientist who uses statistical methods, this book will deepen readers' understanding of why and when methods work and explain how to develop suitable methods for non-standard situations, such as in ecology, big data analysis and genomics.

SIMON N. WOOD works as a professor of statistics at the University of Bath, with particular interests in statistical computing, methodology of smoothing and environmental statistics.

INSTITUTE OF MATHEMATICAL STATISTICS TEXTBOOKS

IMS Textbooks give introductory accounts of topics of current concern suitable for advanced courses at master's level, for doctoral students and for individual study. They are typically shorter than a fully developed textbook, often arising from material created for a topical course. Lengths of 100–290 pages are envisaged. The books typically contain exercises.

Other Books in the Series

1. *Probability on Graphs*, by Geoffrey Grimmett
2. *Stochastic Networks*, by Frank Kelly and Elena Yudovina
3. *Bayesian Filtering and Smoothing*, by Simo Särkkä
4. *The Surprising Mathematics of Longest Increasing Subsequences*, by Dan Romik
5. *Noise Sensitivity of Boolean Functions and Percolation*, by Christophe Garban and Jeffrey E. Steif

Core Statistics

SIMON N. WOOD
University of Bath

Shaftesbury Road, Cambridge CB2 8EA, United Kingdom

One Liberty Plaza, 20th Floor, New York, NY 10006, USA

477 Williamstown Road, Port Melbourne, VIC 3207, Australia

314–321, 3rd Floor, Plot 3, Splendor Forum, Jasola District Centre, New Delhi – 110025, India

103 Penang Road, #05–06/07, Visioncrest Commercial, Singapore 238467

Cambridge University Press is part of Cambridge University Press & Assessment,
a department of the University of Cambridge.

We share the University's mission to contribute to society through the pursuit of
education, learning and research at the highest international levels of excellence.

www.cambridge.org
Information on this title: www.cambridge.org/9781107415041

First published 2015

A catalogue record for this publication is available from the British Library

Library of Congress Cataloging-in-Publication data
Wood, Simon N. Core statistics / Simon Wood, University of Bath.
pages cm
Includes bibliographical references and index.
ISBN 978-1-107-07105-6 (hardback)
1. Statistics – Textbooks. 2. Statistics – Study and teaching (Graduate) I. Title.
QA276.2.W66 2015
519.5–dc23 2014039012

ISBN 978-1-107-07105-6 Hardback
ISBN 978-1-107-41504-1 Paperback

Contents

Preface

This book is aimed at the numerate reader who has probably taken an introductory statistics and probability course at some stage and would like a brief introduction to the core methods of statistics and how they are applied, not necessarily in the context of standard models. The first chapter is a brief review of some basic probability theory needed for what follows. Chapter 2 discusses statistical models and the questions addressed by statistical inference and introduces the maximum likelihood and Bayesian approaches to answering them. Chapter 3 is a short overview of the R programming language. Chapter 4 provides a concise coverage of the large sample theory of maximum likelihood estimation, and Chapter 5 discusses the numerical methods required to use this theory. Chapter 6 covers the numerical methods useful for Bayesian computation, in particular Markov chain Monte Carlo. Chapter 7 provides a brief tour of the theory and practice of linear modelling. Appendices then cover some useful information on common distributions, matrix computation and random number generation. The book is neither an encyclopedia nor a cookbook, and the bibliography aims to provide a compact list of the most useful sources for further reading, rather than being extensive. The aim is to offer a concise coverage of the core knowledge needed to understand and use parametric statistical methods and to build new methods for analysing data. Modern statistics exists at the interface between computation and theory, and this book reflects that fact. I am grateful to Nicole Augustin, Finn Lindgren, the editors at Cambridge University Press and the students on the Bath course 'Applied Statistical Inference' and the Academy for PhD Training in Statistics course 'Statistical Computing' for many useful comments.

1

Random variables

1.1 Random variables

Statistics is about extracting information from data that contain an inherently unpredictable component. *Random variables* are the mathematical construct used to build models of such variability. A random variable takes a different value, at random, each time it is observed. We cannot say, in advance, exactly what value will be taken, but we can make probability statements about the values likely to occur. That is, we can characterise the *distribution* of values taken by a random variable. This chapter briefly reviews the technical constructs used for working with random variables, as well as a number of generally useful related results. See De Groot and Schervish (2002) or Grimmett and Stirzaker (2001) for fuller introductions.

1.2 Cumulative distribution functions

The cumulative distribution function (c.d.f.) of a random variable (r.v.), X, is the function $F(x)$ such that

$$F(x) = \text{Pr}(X \leq x).$$

That is, $F(x)$ gives the probability that the value of X will be less than or equal to x. Obviously, $F(-\infty) = 0$, $F(\infty) = 1$ and $F(x)$ is monotonic. A useful consequence of this definition is that if F is continuous then $F(X)$ has a uniform distribution on $[0, 1]$: it takes any value between 0 and 1 with equal probability. This is because

$$\text{Pr}(X \leq x) = \text{Pr}\{F(X) \leq F(x)\} = F(x) \Rightarrow \text{Pr}\{F(X) \leq u\} = u$$

(if F is continuous), the latter being the c.d.f. of a uniform r.v. on $[0, 1]$.

Define the inverse of the c.d.f. as $F^{-}(u) = \min(x|F(x) \geq u)$, which is just the usual inverse function of F if F is continuous. F^{-} is often called the *quantile function* of X. If U has a uniform distribution on $[0, 1]$, then

$F^-(U)$ is distributed as X with c.d.f. F. Given some way of generating uniform random deviates, this provides a method for generating random variables from any distribution with a computable F^-.

Let p be a number between 0 and 1. The p *quantile* of X is the value that X will be less than or equal to, with probability p. That is, $F^-(p)$. Quantiles have many uses. One is to check whether data, $x_1, x_2, \ldots, x_n$, could plausibly be observations of a random variable with c.d.f. F. The x_i are sorted into order, so that they can be treated as 'observed quantiles'. They are then plotted against the theoretical quantiles $F^-\{(i - 0.5)/n\}$ $(i = 1, \ldots, n)$ to produce a *quantile-quantile plot* (QQ-plot). An approximately straight-line QQ-plot should result, if the observations are from a distribution with c.d.f. F.

1.3 Probability (density) functions

For many statistical methods a function that tells us about the probability of a random value taking a particular value is more useful than the c.d.f. To discuss such functions requires some distinction to be made between random variables taking a discrete set of values (e.g. the non-negative integers) and those taking values from intervals on the real line.

For a discrete random variable, X, the *probability function* (or probability *mass* function), $f(x)$, is the function such that

$$f(x) = \Pr(X = x).$$

Clearly $0 \leq f(x) \leq 1$, and since X must take some value, $\sum_i f(x_i) = 1$, where the summation is over all possible values of x (denoted x_i).

Because a continuous random variable, X, can take an infinite number of possible values, the probability of taking any particular value is usually zero, so that a probability function would not be very useful. Instead the *probability density function*, $f(x)$, gives the probability per unit interval of X being near x. That is, $\Pr(x - \Delta/2 < X < x + \Delta/2) \simeq f(x)\Delta$. More formally, for any constants $a \leq b$,

$$\Pr(a \leq X \leq b) = \int_a^b f(x)dx.$$

Clearly this only works if $f(x) \geq 0$ and $\int_{-\infty}^{\infty} f(x)dx = 1$. Note that $\int_{-\infty}^{b} f(x)dx = F(b)$, so $F'(x) = f(x)$ when F' exists. Appendix A provides some examples of useful standard distributions and their probability (density) functions.

The following sections mostly consider continuous random variables, but except where noted, equivalent results also apply to discrete random variables upon replacement of integration by an appropriate summation. For conciseness the convention is adopted that p.d.f.s with different arguments usually denote different functions (e.g. $f(y)$ and $f(x)$ denote different p.d.f.s).

1.4 Random vectors

Little can usually be learned from single observations. Useful statistical analysis requires multiple observations and the ability to deal simultaneously with multiple random variables. A multivariate version of the p.d.f. is required. The two-dimensional case suffices to illustrate most of the required concepts, so consider random variables X and Y.

The *joint probability density function* of X and Y is the function $f(x, y)$ such that, if Ω is any region in the $x - y$ plane,

$$\Pr\{(X, Y) \in \Omega\} = \iint_{\Omega} f(x, y) dx dy. \tag{1.1}$$

So $f(x, y)$ is the probability *per unit area* of the $x - y$ plane, at x, y. If ω is a small region of area α, containing a point x, y, then $\Pr\{(X, Y) \in \omega\} \simeq f_{xy}(x, y)\alpha$. As with the univariate p.d.f. $f(x, y)$ is non-negative and integrates to one over $\mathbb{R}^2$.

Example Figure 1.1 illustrates the following joint p.d.f.

$$f(x, y) = \begin{cases} x + 3y^2/2 & 0 < x < 1 \,\&\, 0 < y < 1 \\ 0 & \text{otherwise.} \end{cases} \tag{1.2}$$

Figure 1.2 illustrates evaluation of two probabilities using this p.d.f.

1.4.1 Marginal distribution

Continuing with the X, Y case, the p.d.f. of X or Y, ignoring the other variable, can be obtained from $f(x, y)$. To find the *marginal* p.d.f. of X, we seek the probability density of X given that $-\infty < Y < \infty$. From the defining property of a p.d.f., it is unsurprising that this is

$$f(x) = \int_{-\infty}^{\infty} f(x, y) dy,$$

with a similar definition for $f(y)$.

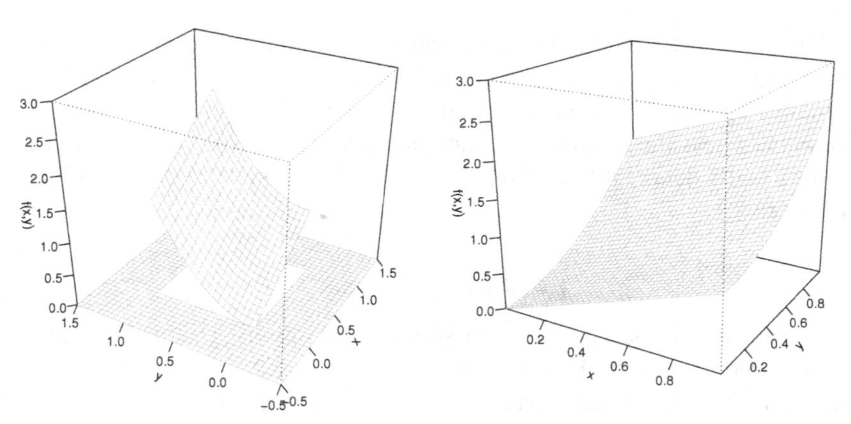

Figure 1.1 The example p.d.f (1.2). Left: over the region $[-0.5, 1.5] \times [-0.5, 1.5]$. Right: the nonzero part of the p.d.f.

1.4.2 Conditional distribution

Suppose that we know that Y takes some particular value y_0. What does this tell us about the distribution of X? Because X and Y have joint density $f(x, y)$, we would expect the density of x, given $Y = y_0$, to be proportional to $f(x, y_0)$. That is, we expect

$$f(x|Y = y_0) = kf(x, y_0),$$

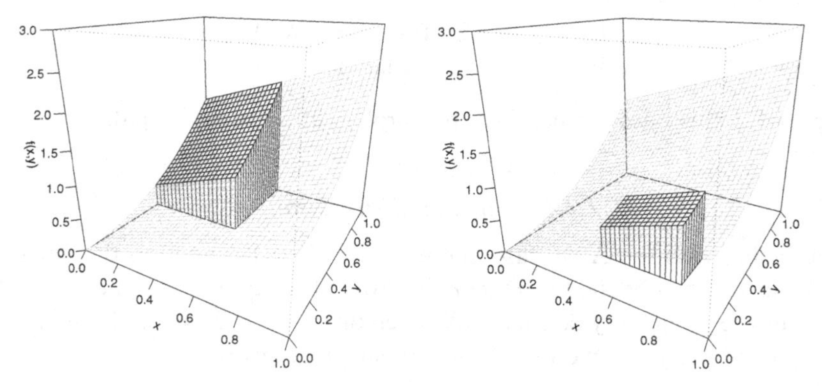

Figure 1.2 Evaluating probabilities from the joint p.d.f. (1.2), shown in grey. Left: in black is shown the volume evaluated to find $\Pr[X < .5, Y > .5]$. Right: $\Pr[.4 < X < .8, .2 < Y < .4]$.

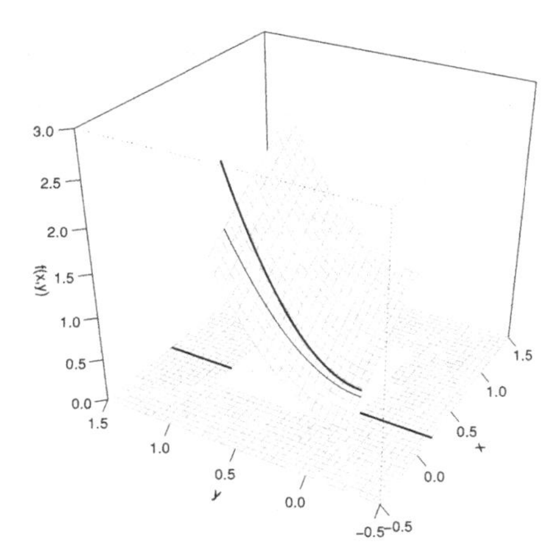

Figure 1.3 The conditional density $f(y|.2)$. The joint density $f(x, y)$ is shown as a grey surface. The thin black curve shows $f(.2, y)$. The thick black curve shows $f(y|.2) = f(.2, y)/f_x(.2)$.

where k is a constant. Now if $f(x|y)$ is a probability density function, then it must integrate to 1. So,

$$k \int_{-\infty}^{\infty} f(x, y_0) dx = 1 \;\Rightarrow\; kf(y_0) = 1 \Rightarrow k = \frac{1}{f(y_0)},$$

where $f(y_0)$ denotes the marginal density of y at y_0. Hence we have:

Definition If X and Y have joint density $f(x, y)$ then the *conditional density* of X, given $Y = y_0$, is

$$f(x|Y = y_0) = \frac{f(x, y_0)}{f(y_0)}, \tag{1.3}$$

assuming $f(y_0) > 0$.

Notice that this is a p.d.f. for random variable X: y_0 is now fixed. To simplify notation we can also write $f(x|y_0)$ in place of $f(x|Y = y_0)$, when the meaning is clear. Of course, symmetric definitions apply to the conditional distribution of Y given X: $f(y|x_0) = f(x_0, y)/f(x_0)$. Figure 1.3 illustrates the relationship between joint and conditional p.d.f.s.

Manipulations involving the replacement of joint distributions with conditional distributions, using $f(x, y) = f(x|y)f(y)$, are common in statistics, but not everything about generalising beyond two dimensions is completely obvious, so the following three examples may help.

1. $f(x, z|y) = f(x|z, y)f(z|y)$.
2. $f(x, z, y) = f(x|z, y)f(z|y)f(y)$.
3. $f(x, z, y) = f(x|z, y)f(z, y)$.

1.4.3 Bayes theorem

From the previous section it is clear that

$$f(x, y) = f(x|y)f(y) = f(y|x)f(x).$$

Rearranging the last two terms gives

$$f(x|y) = \frac{f(y|x)f(x)}{f(y)}.$$

This important result, *Bayes theorem*, leads to a whole school of statistical modelling, as we see in chapters 2 and 6.

1.4.4 Independence and conditional independence

If random variables X and Y are such that $f(x|y)$ does not depend on the value of y, then x is statistically *independent* of y. This has the consequence that

$$\begin{aligned} f(x) = \int_{-\infty}^{\infty} f(x, y)dy &= \int_{-\infty}^{\infty} f(x|y)f(y)dy \\ &= f(x|y) \int_{-\infty}^{\infty} f(y)dy = f(x|y), \end{aligned}$$

which in turn implies that $f(x, y) = f(x|y)f(y) = f(x)f(y)$. Clearly the reverse implication also holds, since $f(x, y) = f(x)f(y)$ leads to $f(x|y) = f(x, y)/f(y) = f(x)f(y)/f(y) = f(x)$. In general then:

Random variables X and Y are independent if and only if their joint p.(d.)f. is given by the product of their marginal p.(d.)f.s: that is, $f(x, y) = f(x)f(y)$.

Modelling the elements of a random vector as independent usually simplifies statistical inference. Assuming independent *identically distributed* (i.i.d.) elements is even simpler, but much less widely applicable.

In many applications, a set of observations cannot be modelled as independent, but can be modelled as *conditionally independent*. Much of modern statistical research is devoted to developing useful models that exploit various sorts of conditional independence in order to model dependent data in computationally feasible ways.

Consider a sequence of random variables $X_1, X_2, \ldots X_n$, and let $\mathbf{X}_{-i} = (X_1, \ldots, X_{i-1}, X_{i+1}, \ldots, X_n)^{\mathrm{T}}$. A simple form of conditional independence is the first order Markov property,

$$f(x_i|\mathbf{x}_{-i}) = f(x_i|x_{i-1}).$$

That is, X_{i-1} completely determines the distribution of X_i, so that *given* X_{i-1}, X_i is independent of the rest of the sequence. It follows that

$$\begin{aligned} f(\mathbf{x}) = f(x_n|\mathbf{x}_{-n})f(\mathbf{x}_{-n}) &= f(x_n|x_{n-1})f(\mathbf{x}_{-n}) \\ &= \ldots = \prod_{i=2}^{n} f(x_i|x_{i-1})f(x_1), \end{aligned}$$

which can often be exploited to yield considerable computational savings.

1.5 Mean and variance

Although it is important to know how to characterise the distribution of a random variable completely, for many purposes its first- and second-order properties suffice. In particular the *mean* or *expected value* of a random variable, X, with p.d.f. $f(x)$, is defined as

$$E(X) = \int_{-\infty}^{\infty} x f(x) dx.$$

Since the integral is weighting each possible value of x by its relative frequency of occurrence, we can interpret $E(X)$ as being the average of an infinite sequence of observations of X.

The definition of expectation applies to any function g of X:

$$E\{g(X)\} = \int_{-\infty}^{\infty} g(x) f(x) dx.$$

Defining $\mu = E(X)$, then a particularly useful g is $(X - \mu)^2$, measuring the squared difference between X and its average value, which is used to define the *variance* of X:

$$\text{var}(X) = E\{(X - \mu)^2\}.$$

The variance of X measures how spread out the distribution of X is. Although computationally convenient, its interpretability is hampered by having units that are the square of the units of X. The *standard deviation* is the square root of the variance, and hence is on the same scale as X.

1.5.1 Mean and variance of linear transformations

From the definition of expectation it follows immediately that if a and b are finite real constants $E(a + bX) = a + bE(X)$. The variance of $a + bX$ requires slightly more work:

$$\begin{aligned}\text{var}(a + bX) &= E\{(a + bX - a - b\mu)^2\}\\ &= E\{b^2(X - \mu)^2\} = b^2E\{(X - \mu)^2\} = b^2\text{var}(X).\end{aligned}$$

If X and Y are random variables then $E(X + Y) = E(X) + E(Y)$. To see this suppose that they have joint density $f(x, y)$; then,

$$\begin{aligned}E(X + Y) &= \int (x + y)f(x, y)dxdy\\ &= \int xf(x, y)dxdy + \int yf(x, y)dxdy = E(X) + E(Y).\end{aligned}$$

This result assumes nothing about the distribution of X and Y. If we now add the assumption that X and Y are independent then we find that $E(XY) = E(X)E(Y)$ as follows:

$$\begin{aligned}E(XY) &= \int xyf(x, y)dxdy\\ &= \int xf(x)yf(y)dxdy \text{ (by independence)}\\ &= \int xf(x)dx \int yf(y)dy = E(X)E(Y).\end{aligned}$$

Note that the reverse implication only holds if the joint distribution of X and Y is Gaussian.

Variances do not add as nicely as means (unless X and Y are independent), and we need the notion of *covariance*:

$$\text{cov}(X, Y) = E\{(X - \mu_x)(Y - \mu_y)\} = E(XY) - E(X)E(Y),$$

where $\mu_x = E(X)$ and $\mu_y = E(Y)$. Clearly $\text{var}(X) \equiv \text{cov}(X, X)$, and if X and Y are independent $\text{cov}(X, Y) = 0$ (since then $E(XY) = E(X)E(Y)$).

Now let $\mathbf{A}$ and $\mathbf{b}$ be, respectively, a matrix and a vector of fixed finite coefficients, with the same number of rows, and let $\mathbf{X}$ be a random vector. $E(\mathbf{X}) = \boldsymbol{\mu}_x = \{E(X_1), E(X_2), \ldots, E(X_n)\}^{\mathrm{T}}$ and it is immediate that $E(\mathbf{AX} + \mathbf{b}) = \mathbf{A}E(\mathbf{X}) + \mathbf{b}$. A useful summary of the second-order properties of $\mathbf{X}$ requires both variances and covariances of its elements. These can be written in the (symmetric) variance-covariance matrix $\boldsymbol{\Sigma}$, where $\Sigma_{ij} = \mathrm{cov}(X_i, X_j)$, which means that

$$\boldsymbol{\Sigma} = E\{(\mathbf{X} - \boldsymbol{\mu}_x)(\mathbf{X} - \boldsymbol{\mu}_x)^{\mathrm{T}}\}. \tag{1.4}$$

A very useful result is that

$$\boldsymbol{\Sigma}_{AX+b} = \mathbf{A}\boldsymbol{\Sigma}\mathbf{A}^{\mathrm{T}}, \tag{1.5}$$

which is easily proven:

$$\begin{aligned}
\boldsymbol{\Sigma}_{AX+b} &= E\{(\mathbf{AX} + \mathbf{b} - \mathbf{A}\boldsymbol{\mu}_x - \mathbf{b})(\mathbf{AX} + \mathbf{b} - \mathbf{A}\boldsymbol{\mu}_x - \mathbf{b})^{\mathrm{T}}\} \\
&= E\{(\mathbf{AX} - \mathbf{A}\boldsymbol{\mu}_x)(\mathbf{AX} - \mathbf{A}\boldsymbol{\mu}_x)^{\mathrm{T}}) \\
&= \mathbf{A}E\{(\mathbf{X} - \boldsymbol{\mu}_x)(\mathbf{X} - \boldsymbol{\mu}_x)^{\mathrm{T}}\}\mathbf{A}^{\mathrm{T}} = \mathbf{A}\boldsymbol{\Sigma}\mathbf{A}^{\mathrm{T}}.
\end{aligned}$$

So if $\mathbf{a}$ is a vector of fixed real coefficients then $\mathrm{var}(\mathbf{a}^{\mathrm{T}}\mathbf{X}) = \mathbf{a}^{\mathrm{T}}\boldsymbol{\Sigma}\mathbf{a} \geq 0$: a covariance matrix is positive semi-definite.

1.6 The multivariate normal distribution

The normal or *Gaussian* distribution (see Section A.1.1) has a central place in statistics, largely as a result of the central limit theorem covered in Section 1.9. Its multivariate version is particularly useful.

Definition Consider a set of n i.i.d. standard normal random variables: $Z_i \underset{\text{i.i.d}}{\sim} N(0, 1)$. The covariance matrix for $\mathbf{Z}$ is $\mathbf{I}_n$ and $E(\mathbf{Z}) = \mathbf{0}$. Let $\mathbf{B}$ be an $m \times n$ matrix of fixed finite real coefficients and $\boldsymbol{\mu}$ be an m- vector of fixed finite real coefficients. The m-vector $\mathbf{X} = \mathbf{BZ} + \boldsymbol{\mu}$ is said to have a *multivariate normal distribution*. $E(\mathbf{X}) = \boldsymbol{\mu}$ and the covariance matrix of $\mathbf{X}$ is just $\boldsymbol{\Sigma} = \mathbf{BB}^{\mathrm{T}}$. The short way of writing $\mathbf{X}$'s distribution is

$$\mathbf{X} \sim N(\boldsymbol{\mu}, \boldsymbol{\Sigma}).$$

In Section 1.7, basic transformation theory establishes that the p.d.f. for this distribution is

$$f_{\mathbf{x}}(\mathbf{x}) = \frac{1}{\sqrt{(2\pi)^m |\boldsymbol{\Sigma}|}} e^{-\frac{1}{2}(\mathbf{x}-\boldsymbol{\mu})^{\mathrm{T}}\boldsymbol{\Sigma}^{-1}(\mathbf{x}-\boldsymbol{\mu})} \quad \text{for} \quad \mathbf{x} \in \mathbb{R}^m, \tag{1.6}$$

assuming $\mathbf{\Sigma}$ has full rank (if $m = 1$ the definition gives the usual univariate normal p.d.f.). Actually there exists a more general definition in which $\mathbf{\Sigma}$ is merely positive semi-definite, and hence potentially singular: this involves a *pseudoinverse* of $\mathbf{\Sigma}$.

An interesting property of the multivariate normal distribution is that if X and Y have a multivariate normal distribution and zero covariance, then they must be independent. This implication only holds for the normal (independence implies zero covariance for any distribution).

1.6.1 A multivariate t distribution

If we replace the random variables $Z_i \underset{\text{i.i.d}}{\sim} N(0,1)$ with random variables $T_i \underset{\text{i.i.d}}{\sim} t_k$ (see Section A.1.3) in the definition of a multivariate normal, we obtain a vector with a multivariate $t_k(\boldsymbol{\mu}, \mathbf{\Sigma})$ distribution. This can be useful in stochastic simulation, when we need a multivariate distribution with heavier tails than the multivariate normal. Note that the resulting univariate marginal distributions are not t distributed. Multivariate t densities with t distributed marginals are more complicated to characterise.

1.6.2 Linear transformations of normal random vectors

From the definition of multivariate normality, it immediately follows that if $\mathbf{X} \sim N(\boldsymbol{\mu}, \mathbf{\Sigma})$ and $\mathbf{A}$ is a matrix of finite real constants (of suitable dimensions), then

$$\mathbf{AX} \sim N(\mathbf{A}\boldsymbol{\mu}, \mathbf{A\Sigma A}^{\mathrm{T}}). \tag{1.7}$$

This is because $\mathbf{X} = \mathbf{BZ} + \boldsymbol{\mu}$, so $\mathbf{AX} = \mathbf{ABZ} + \mathbf{A}\boldsymbol{\mu}$, and hence $\mathbf{AX}$ is exactly the sort of linear transformation of standard normal r.v.s that defines a multivariate normal random vector. Furthermore it is clear that $E(\mathbf{AX}) = \mathbf{A}\boldsymbol{\mu}$ and the covariance matrix of $\mathbf{AX}$ is $\mathbf{A\Sigma A}^{\mathrm{T}}$.

A special case is that if $\mathbf{a}$ is a vector of finite real constants, then

$$\mathbf{a}^{\mathrm{T}}\mathbf{X} \sim N(\mathbf{a}^{\mathrm{T}}\boldsymbol{\mu}, \mathbf{a}^{\mathrm{T}}\mathbf{\Sigma a}).$$

For the case in which $\mathbf{a}$ is a vector of zeros, except for a_j, which is 1, (1.7) implies that

$$X_j \sim N(\mu_j, \Sigma_{jj}) \tag{1.8}$$

(usually we would write σ_j^2 for Σ_{jj}). In words:

If $\mathbf{X}$ has a multivariate normal distribution, then the marginal distribution of any X_j is univariate normal.

More generally, the marginal distribution of any subvector of $\mathbf{X}$ is multivariate normal, by a similar argument to that which led to (1.8).

The reverse implication does not hold. Marginal normality of the X_j is **not** sufficient to imply that $\mathbf{X}$ has a multivariate normal distribution. However, if $\mathbf{a}^{\mathrm{T}}\mathbf{X}$ has a normal distribution, for all (finite real) $\mathbf{a}$, then $\mathbf{X}$ must have a multivariate normal distribution.

1.6.3 Multivariate normal conditional distributions

Suppose that $\mathbf{Z}$ and $\mathbf{X}$ are random vectors with a multivariate normal joint distribution. Partitioning their joint covariance matrix

$$\boldsymbol{\Sigma} = \begin{bmatrix} \boldsymbol{\Sigma}_z & \boldsymbol{\Sigma}_{zx} \\ \boldsymbol{\Sigma}_{xz} & \boldsymbol{\Sigma}_x \end{bmatrix},$$

then

$$\mathbf{X}|\mathbf{z} \sim N(\boldsymbol{\mu}_x + \boldsymbol{\Sigma}_{xz}\boldsymbol{\Sigma}_z^{-1}(\mathbf{z} - \boldsymbol{\mu}_z), \boldsymbol{\Sigma}_x - \boldsymbol{\Sigma}_{xz}\boldsymbol{\Sigma}_z^{-1}\boldsymbol{\Sigma}_{zx}).$$

Proof relies on a result for the inverse of a symmetric partitioned matrix:

$$\begin{bmatrix} \mathbf{A} & \mathbf{C} \\ \mathbf{C}^{\mathrm{T}} & \mathbf{B} \end{bmatrix}^{-1} = \begin{bmatrix} \mathbf{A}^{-1} + \mathbf{A}^{-1}\mathbf{C}\mathbf{D}^{-1}\mathbf{C}^{\mathrm{T}}\mathbf{A}^{-1} & -\mathbf{A}^{-1}\mathbf{C}\mathbf{D}^{-1} \\ -\mathbf{D}^{-1}\mathbf{C}^{\mathrm{T}}\mathbf{A}^{-1} & \mathbf{D}^{-1} \end{bmatrix}$$

where $\mathbf{D} = \mathbf{B} - \mathbf{C}^{\mathrm{T}}\mathbf{A}^{-1}\mathbf{C}$ (this can be checked easily, if tediously). Now find the conditional p.d.f. of $\mathbf{X}$ given $\mathbf{Z}$. Defining $\mathbf{Q} = \boldsymbol{\Sigma}_x - \boldsymbol{\Sigma}_{xz}\boldsymbol{\Sigma}_z^{-1}\boldsymbol{\Sigma}_{zx}$, $\tilde{\mathbf{z}} = \mathbf{z} - \boldsymbol{\mu}_z$, $\tilde{\mathbf{x}} = \mathbf{x} - \boldsymbol{\mu}_x$ and noting that terms involving only z are part of the normalising constant,

$$\begin{aligned} f(\mathbf{x}|\mathbf{z}) &= f(\mathbf{x},\mathbf{z})/f(\mathbf{z}) \\ &\propto \exp\left\{-\frac{1}{2}\begin{bmatrix} \tilde{\mathbf{z}} \\ \tilde{\mathbf{x}} \end{bmatrix}^{\mathrm{T}} \begin{bmatrix} \boldsymbol{\Sigma}_z^{-1} + \boldsymbol{\Sigma}_z^{-1}\boldsymbol{\Sigma}_{zx}\mathbf{Q}^{-1}\boldsymbol{\Sigma}_{xz}\boldsymbol{\Sigma}_z^{-1} & -\boldsymbol{\Sigma}_z^{-1}\boldsymbol{\Sigma}_{zx}\mathbf{Q}^{-1} \\ -\mathbf{Q}^{-1}\boldsymbol{\Sigma}_{xz}\boldsymbol{\Sigma}_z^{-1} & \mathbf{Q}^{-1} \end{bmatrix} \begin{bmatrix} \tilde{\mathbf{z}} \\ \tilde{\mathbf{x}} \end{bmatrix}\right\} \\ &\propto \exp\left\{-\tilde{\mathbf{x}}^{\mathrm{T}}\mathbf{Q}^{-1}\tilde{\mathbf{x}}/2 + \tilde{\mathbf{x}}^{\mathrm{T}}\mathbf{Q}^{-1}\boldsymbol{\Sigma}_{xz}\boldsymbol{\Sigma}_z^{-1}\tilde{\mathbf{z}} + z \text{ terms}\right\} \\ &\propto \exp\left\{-(\tilde{\mathbf{x}} - \boldsymbol{\Sigma}_{xz}\boldsymbol{\Sigma}_z^{-1}\tilde{\mathbf{z}})^{\mathrm{T}}\mathbf{Q}^{-1}(\tilde{\mathbf{x}} - \boldsymbol{\Sigma}_{xz}\boldsymbol{\Sigma}_z^{-1}\tilde{\mathbf{z}})/2 + z \text{ terms}\right\}, \end{aligned}$$

which is recognisable as a $N(\boldsymbol{\mu}_x + \boldsymbol{\Sigma}_{xz}\boldsymbol{\Sigma}_z^{-1}(\mathbf{z} - \boldsymbol{\mu}_z), \boldsymbol{\Sigma}_x - \boldsymbol{\Sigma}_{xz}\boldsymbol{\Sigma}_z^{-1}\boldsymbol{\Sigma}_{zx})$ p.d.f.

1.7 Transformation of random variables

Consider a continuous random variable Z, with p.d.f. f_z. Suppose $X = g(Z)$ where g is an invertible function. The c.d.f of X is easily obtained from that of Z:

$$
\begin{aligned}
F_x(x) &= \Pr(X \le x) \\
&= \begin{cases} \Pr\{g^{-1}(X) \le g^{-1}(x)\} = \Pr\{Z \le g^{-1}(x)\}, & g \text{ increasing} \\ \Pr\{g^{-1}(X) > g^{-1}(x)\} = \Pr\{Z > g^{-1}(x)\}, & g \text{ decreasing} \end{cases} \\
&= \begin{cases} F_z\{g^{-1}(x)\}, & g \text{ increasing} \\ 1 - F_z\{g^{-1}(x)\}, & g \text{ decreasing} \end{cases}
\end{aligned}
$$

To obtain the p.d.f. we simply differentiate and, whether g is increasing or decreasing, obtain

$$f_x(x) = F_x'(x) = F_z'\{g^{-1}(x)\}\left|\frac{\mathrm{d}z}{\mathrm{d}x}\right| = f_z\{g^{-1}(x)\}\left|\frac{\mathrm{d}z}{\mathrm{d}x}\right|.$$

If g is a vector function and $\mathbf{Z}$ and $\mathbf{X}$ are vectors of the same dimension, then this last result generalises to

$$f_x(\mathbf{x}) = f_z\{g^{-1}(\mathbf{x})\}\,|\mathbf{J}|,$$

where $J_{ij} = \partial z_i/\partial x_j$ (again a one-to-one mapping between $\mathbf{x}$ and $\mathbf{z}$ is assumed). Note that if f_x and f_z are probability functions for discrete random variables then no $|\mathbf{J}|$ term is needed.

Example Use the definition of a multivariate normal random vector to obtain its p.d.f. Let $\mathbf{X} = \mathbf{BZ} + \boldsymbol{\mu}$, where $\mathbf{B}$ is an $n \times n$ invertible matrix and $\mathbf{Z}$ a vector of i.i.d. standard normal random variables. So the covariance matrix of $\mathbf{X}$ is $\boldsymbol{\Sigma} = \mathbf{BB}^{\mathrm{T}}$, $\mathbf{Z} = \mathbf{B}^{-1}(\mathbf{X} - \boldsymbol{\mu})$ and the Jacobian here is $|\mathbf{J}| = |\mathbf{B}^{-1}|$. Since the Z_i are i.i.d. their joint density is the product of their marginals, i.e.

$$f(\mathbf{z}) = \frac{1}{\sqrt{2\pi}^n} e^{-\mathbf{z}^{\mathrm{T}}\mathbf{z}/2}.$$

Direct application of the preceding transformation theory then gives

$$
\begin{aligned}
f(\mathbf{x}) &= \frac{|\mathbf{B}^{-1}|}{\sqrt{2\pi}^n} e^{-(\mathbf{x}-\boldsymbol{\mu})^{\mathrm{T}}\mathbf{B}^{-\mathrm{T}}\mathbf{B}^{-1}(\mathbf{x}-\boldsymbol{\mu})/2} \\
&= \frac{1}{\sqrt{(2\pi)^n|\boldsymbol{\Sigma}|}} e^{-(\mathbf{x}-\boldsymbol{\mu})^{\mathrm{T}}\boldsymbol{\Sigma}^{-1}(\mathbf{x}-\boldsymbol{\mu})/2}.
\end{aligned}
$$

1.8 Moment generating functions

Another characterisation of the distribution of a random variable, X, is its *moment generating function* (m.g.f.),

$$M_X(s) = E\left(e^{sX}\right),$$

where s is real. The kth derivative of the m.g.f. evaluated at $s = 0$ is the kth (uncentered) moment of X:

$$\left.\frac{\mathrm{d}^k M_X}{\mathrm{d}s^k}\right|_{s=0} = E(X^k).$$

So $M_X(0) = 1$, $M_X'(0) = E(X)$, $M_X''(0) = E(X^2)$, etc.

The following three properties will be useful in the next section:

1. If $M_X(s) = M_Y(s)$ for some small interval around $s = 0$, then X and Y are identically distributed.
2. If X and Y are independent, then
$$\begin{aligned} M_{X+Y}(s) &= E\left\{e^{s(X+Y)}\right\} = E\left(e^{sX}e^{sY}\right) = E\left(e^{sX}\right)E\left(e^{sY}\right) \\ &= M_X(s)M_Y(s). \end{aligned}$$
3. $M_{a+bX}(s) = E(e^{as+bXs}) = e^{as}M_X(bs)$.

Property 1 is unsurprising, given that the m.g.f. encodes all the moments of X.

1.9 The central limit theorem

Consider i.i.d. random variables, $X_1, X_2, \ldots X_n$, with mean μ and finite variance σ^2. Let $\bar{X}_n = \sum_{i=1}^n X_i/n$. In its simplest form, the central limit theorem says that in the limit $n \to \infty$,

$$\bar{X}_n \sim N(\mu, \sigma^2/n).$$

Intuitively, consider a Taylor expansion of $l(\bar{x}_n) = \log f(\bar{x}_n)$ where f is the unknown p.d.f. of $\bar{X}_n$, with mode $\hat{x}_n$:

$$f(\bar{x}) \simeq \exp\{l(\hat{x}_n) + l''(\bar{x}_n - \hat{x}_n)^2/2 + l'''(\bar{x}_n - \hat{x}_n)^3/6 + \cdots\}$$

as $n \to \infty$, $\bar{x}_n - \hat{x}_n \to 0$, so that the right hand side tends to an $N(\hat{x}, -1/l'')$ p.d.f. This argument is not rigorous, because it makes implicit assumptions about how derivatives of l vary with n.

A proper proof uses moment generating functions. Define

$$Y_i = \frac{X_i - \mu}{\sigma} \quad \text{and} \quad Z_n = \frac{1}{\sqrt{n}} \sum_{i=1}^{n} Y_i = \frac{\bar{X}_n - \mu}{\sigma/\sqrt{n}}.$$

Now express the m.g.f. of Z_n in terms of the Taylor expansion of the m.g.f. of Y_i (noting that $M_Y'(0) = 0$ and $M_Y''(0) = 1$):

$$\begin{aligned} M_{Z_n}(s) &= \left\{M_Y(s/\sqrt{n})\right\}^n \\ &= \left\{M_Y(0) + M_Y'(0)\frac{s}{\sqrt{n}} + M_Y''(0)\frac{s^2}{2n} + o(n^{-1})\right\}^n \\ &= \left\{1 + \frac{s^2}{2n} + o(n^{-1})\right\}^n = \exp\left[n \log\left\{1 + \frac{s^2}{2n} + o(n^{-1})\right\}\right] \\ &\to \exp\left(\frac{s^2}{2}\right) \quad \text{as} \quad n \to \infty. \end{aligned}$$

The final expression is the m.g.f. of $N(0, 1)$, completing the proof.

The central limit theorem generalises to multivariate and non-identical distribution settings. There are also many non-independent situations where a normal limiting distribution occurs. The theorem is important in statistics because it justifies using the normal as an approximation in many situations where a random variable can be viewed as a sum of other random variables. This applies in particular to the distribution of statistical estimators, which very often have normal distributions in the large sample limit.

1.10 Chebyshev, Jensen and the law of large numbers

Some other general results are useful in what follows.

1.10.1 Chebyshev's inequality

If X is a random variable and $E(X^2) < \infty$, then

$$\Pr(|X| \geq a) \leq \frac{E(X^2)}{a^2}. \tag{1.9}$$

Proof: From the definition of expectation we have

$$\begin{aligned} E(X^2) = E(X^2 \mid a \leq |X|)\Pr(a \leq |X|) \\ + E(X^2 \mid a > |X|)\Pr(a > |X|) \end{aligned}$$

and because all the terms on the right hand side are non-negative it follows that $E(X^2) \geq E(X^2 \mid a \leq |X|)\text{Pr}(a \leq |X|)$. However if $a \leq |X|$, then obviously $a^2 \leq E(X^2 \mid a \leq |X|)$ so $E(X^2) \geq a^2\text{Pr}(|X| \geq a)$ and (1.9) is proven.

1.10.2 The law of large numbers

Consider i.i.d. random variables, $X_1, \ldots X_n$, with mean μ, and $E(|X_i|) < \infty$. If $\bar{X}_n = \sum_{i=1}^n X_i/n$ then the *strong law of large numbers* states that, for any positive ϵ

$$\text{Pr}\left(\lim_{n\to\infty} |\bar{X}_n - \mu| < \epsilon\right) = 1$$

(i.e. $\bar{X}_n$ converges *almost surely* to μ).

Adding the assumption $\text{var}(X_i) = \sigma^2 < \infty$, it is easy to prove the slightly weaker result

$$\lim_{n\to\infty} \text{Pr}\left(|\bar{X}_n - \mu| \geq \epsilon\right) = 0,$$

which is the *weak law of large numbers* (X_n converges *in probability* to μ). A proof is as follows:

$$\text{Pr}\left(|\bar{X}_n - \mu| \geq \epsilon\right) \leq \frac{E(\bar{X}_n - \mu)^2}{\epsilon^2} = \frac{\text{var}(\bar{X}_n)}{\epsilon^2} = \frac{\sigma^2}{n\epsilon^2}$$

and the final term tends to 0 as $n \to \infty$. The inequality is Chebyshev's. Note that the i.i.d. assumption has only been used to ensure that $\text{var}(\bar{X}_n) = \sigma^2/n$. All that we actually needed for the proof was the milder assumption that $\lim_{n\to\infty} \text{var}(\bar{X}_n) = 0$.

To some extent the laws of large numbers are almost statements of the obvious. If they did not hold then random variables would not be of much use for building statistical models.

1.10.3 Jensen's inequality

This states that for any random variable X and concave function c,

$$c\{E(X)\} \geq E\{c(X)\}. \tag{1.10}$$

The proof is most straightforward for a discrete random variable. A concave function, c, is one for which

$$c(w_1x_1 + w_2x_2) \geq w_1c(x_1) + w_2c(x_2) \tag{1.11}$$

for any real non-negative w_1 and w_2 such that $w_1 + w_2 = 1$. Now suppose that it is true that

$$c\left(\sum_{i=1}^{n-1} w_i' x_i\right) \geq \sum_{i=1}^{n-1} w_i' c(x_i) \tag{1.12}$$

for any non-negative constants w_i' such that $\sum_{i=1}^{n-1} w_i' = 1$. Consider any set of non-negative constants w_i such that $\sum_{i=1}^{n} w_i = 1$. We can write

$$\begin{aligned} c\left(\sum_{i=1}^{n} w_i x_i\right) &= c\left((1-w_n)\sum_{i=1}^{n-1} \frac{w_i x_i}{1-w_n} + w_n x_n\right) \\ &\geq (1-w_n) c\left(\sum_{i=1}^{n-1} \frac{w_i x_i}{1-w_n}\right) + w_n c(x_n) \end{aligned} \tag{1.13}$$

where the final inequality is by (1.11). Now from $\sum_{i=1}^{n} w_i = 1$ it follows that $\sum_{i=1}^{n-1} w_i/(1-w_n) = 1$, so (1.12) applies and

$$c\left(\sum_{i=1}^{n-1} \frac{w_i x_i}{1-w_n}\right) \geq \sum_{i=1}^{n-1} \frac{w_i c(x_i)}{1-w_n}.$$

Substituting this into the right hand side of (1.13) results in

$$c\left(\sum_{i=1}^{n} w_i x_i\right) \geq \sum_{i=1}^{n} w_i c(x_i). \tag{1.14}$$

For $n = 3$ (1.12) is just (1.11) and is therefore true. It follows, by induction, that (1.14) is true for any n. By setting $w_i = f(x_i)$, where $f(x)$ is the probability function of the r.v. X, (1.10) follows immediately for a discrete random variable. In the case of a continuous random variable we need to replace the expectation integral by the limit of a discrete weighted sum, and (1.10) again follows from (1.14)

1.11 Statistics

A *statistic* is a function of a set of random variables. Statistics are themselves random variables. Obvious examples are the sample mean and sample variance of a set of data, $x_1, x_2, \ldots x_n$:

$$\bar{x} = \frac{1}{n}\sum_{i=1}^{n} x_i, \quad s^2 = \frac{1}{n-1}\sum_{i=1}^{n}(x_i - \bar{x})^2.$$

The fact that formal statistical procedures can be characterised as functions of sets of random variables (data) accounts for the field's name.

If a statistic $t(\mathbf{x})$ (scalar or vector) is such that the p.d.f. of $\mathbf{x}$ can be written as

$$f_\theta(\mathbf{x}) = h(\mathbf{x}) g_\theta\{t(\mathbf{x})\},$$

where h does not depend on $\boldsymbol{\theta}$ and g depends on $\mathbf{x}$ only through $t(\mathbf{x})$, then t is a *sufficient statistic* for $\boldsymbol{\theta}$, meaning that all information about $\boldsymbol{\theta}$ contained in $\mathbf{x}$ is provided by $t(\mathbf{x})$. See Section 4.1 for a formal definition of 'information'. Sufficiency also means that the distribution of $\mathbf{x}$ given $t(\mathbf{x})$ does not depend on $\boldsymbol{\theta}$.

Exercises

1.1 Exponential random variable, $X \geq 0$, has p.d.f. $f(x) = \lambda \exp(-\lambda x)$.

1. Find the c.d.f. and the quantile function for X.
2. Find $\Pr(X < \lambda)$ and the median of X.
3. Find the mean and variance of X.

1.2 Evaluate $\Pr(X < 0.5, Y < 0.5)$ if X and Y have joint p.d.f. (1.2).

1.3 Suppose that

$$\mathbf{Y} \sim N\left(\begin{bmatrix} 1 \\ 2 \end{bmatrix}, \begin{bmatrix} 2 & 1 \\ 1 & 2 \end{bmatrix} \right).$$

Find the conditional p.d.f. of Y_1 given that $Y_1 + Y_2 = 3$.

1.4 If $\mathbf{Y} \sim N(\boldsymbol{\mu}, \mathbf{I}\sigma^2)$ and $\mathbf{Q}$ is any orthogonal matrix of appropriate dimension, find the distribution of $\mathbf{QY}$. Comment on what is surprising about this result.

1.5 If $\mathbf{X}$ and $\mathbf{Y}$ are independent random vectors of the same dimension, with covariance matrices $\mathbf{V}_x$ and $\mathbf{V}_y$, find the covariance matrix of $\mathbf{X} + \mathbf{Y}$.

1.6 Let X and Y be non-independent random variables, such that $\text{var}(X) = \sigma_x^2$, $\text{var}(Y) = \sigma_y^2$ and $\text{cov}(X, Y) = \sigma_{xy}^2$. Using the result from Section 1.6.2, find $\text{var}(X + Y)$ and $\text{var}(X - Y)$.

1.7 Let Y_1, Y_2 and Y_3 be independent $N(\mu, \sigma^2)$ r.v.s. Somehow using the matrix

$$\begin{bmatrix} 1/3 & 1/3 & 1/3 \\ 2/3 & -1/3 & -1/3 \\ -1/3 & 2/3 & -1/3 \end{bmatrix}$$

show that $\bar{Y} = \sum_{i=1}^3 Y_i/3$ and $\sum_{i=1}^3 (Y_i - \bar{Y})^2$ are independent random variables.

1.8 If $\log(X) \sim N(\mu, \sigma^2)$, find the p.d.f. of X.

1.9 Discrete random variable Y has a Poisson distribution with parameter λ if its p.d.f. is $f(y) = \lambda^y e^{-\lambda}/y!$, for $y = 0, 1, \ldots$

a. Find the moment generating function for Y (hint: the power series representation of the exponential function is useful).
b. If $Y_1 \sim \text{Poi}(\lambda_1)$ and independently $Y_2 \sim \text{Poi}(\lambda_2)$, deduce the distribution of $Y_1 + Y_2$, by employing a general property of m.g.f.s.
c. Making use of the previous result and the central limit theorem, deduce the normal approximation to the Poisson distribution.
d. Confirm the previous result graphically, using R functions `dpois`, `dnorm`, `plot` or `barplot` and `lines`. Confirm that the approximation improves with increasing λ.

2

Statistical models and inference

Statistics aims to extract information from data: specifically, information about the system that generated the data. There are two difficulties with this enterprise. First, it may not be easy to infer what we want to know from the data that can be obtained. Second, most data contain a component of random variability: if we were to replicate the data-gathering process several times we would obtain somewhat different data on each occasion. In the face of such variability, how do we ensure that the conclusions drawn from a single set of data are generally valid, and not a misleading reflection of the random peculiarities of that single set of data?

Statistics provides methods for overcoming these difficulties and making sound inferences from inherently random data. For the most part this involves the use of *statistical models*, which are like 'mathematical cartoons' describing how our data might have been generated, if the unknown features of the data-generating system were actually known. So if the unknowns were known, then a decent model could generate data that resembled the observed data, including reproducing its variability under replication. The purpose of *statistical inference* is then to use the statistical model to go in the reverse direction: to infer the values of the model unknowns that are consistent with observed data.

Mathematically, let $\mathbf{y}$ denote a random vector containing the observed data. Let $\boldsymbol{\theta}$ denote a vector of parameters of unknown value. We assume that knowing the values of some of these parameters would answer the questions of interest about the system generating $\mathbf{y}$. So a statistical model is a recipe by which $\mathbf{y}$ might have been generated, given appropriate values for $\boldsymbol{\theta}$. At a minimum the model specifies how data like $\mathbf{y}$ might be simulated, thereby implicitly defining the distribution of $\mathbf{y}$ and how it depends on $\boldsymbol{\theta}$. Often it will provide more, by explicitly defining the p.d.f. of $\mathbf{y}$ in terms of $\boldsymbol{\theta}$. Generally a statistical model may also depend on some known parameters, $\boldsymbol{\gamma}$, and some further data, $\mathbf{x}$, that are treated as known and are referred to as *covariates* or *predictor variables*. See Figure 2.1.

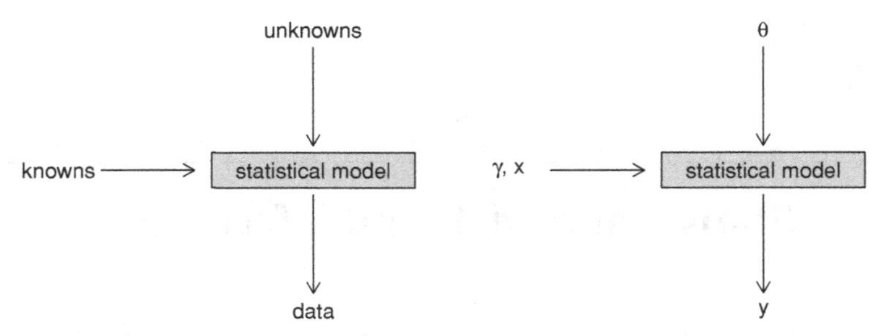

Figure 2.1 Left: a statistical model is a mathematical description of how the values of some knowns and unknowns could have been used to generate observed data and other stochastically similar replicates. Right: the model unknowns are written in a parameter vector $\boldsymbol{\theta}$ and the model knowns may include fixed data, $\mathbf{x}$ and parameters $\boldsymbol{\gamma}$. The data are observations, $\mathbf{y}$, of a random vector. At minimum a statistical model must allow random data to be simulated that are stochastically similar to $\mathbf{y}$: explicitly or implicitly it specifies the distribution of $\mathbf{y}$ in terms of $\mathbf{x}$, $\boldsymbol{\gamma}$ and $\boldsymbol{\theta}$. Statistical methods aim to reverse the direction of the vertical arrows: to infer the unknown $\boldsymbol{\theta}$ from the observed data $\mathbf{y}$.

In short, if we knew the value of $\boldsymbol{\theta}$, a correct statistical model would allow us to simulate as many replicate random data vectors $\mathbf{y}^*$ as we like, which should all resemble our observed data $\mathbf{y}$ (while almost never being identical to it). Statistical methods are about taking models specified in this unknown parameters to known data way and automatically inverting them to work out the values of the unknown parameters $\boldsymbol{\theta}$ that are consistent with the known observed data $\mathbf{y}$.

2.1 Examples of simple statistical models

1. Consider the following 60-year record of mean annual temperatures in New Haven, Connecticut (in °F, and available as `nhtemp` in R).

```
49.9 52.3 49.4 51.1 49.4 47.9 49.8 50.9 49.3 51.9 50.8 49.6 49.3 50.6
48.4 50.7 50.9 50.6 51.5 52.8 51.8 51.1 49.8 50.2 50.4 51.6 51.8 50.9
48.8 51.7 51.0 50.6 51.7 51.5 52.1 51.3 51.0 54.0 51.4 52.7 53.1 54.6
52.0 52.0 50.9 52.6 50.2 52.6 51.6 51.9 50.5 50.9 51.7 51.4 51.7 50.8
51.9 51.8 51.9 53.0
```

A simple model would treat these data as independent observations from an $N(\mu, \sigma^2)$ distribution, where μ and σ^2 are unknown parameters (see Section A.1.1). Then the p.d.f. for the random variable corresponding to a single measurement, y_i, is

$$f(y_i) = \frac{1}{\sqrt{2\pi}\sigma} e^{\frac{-(y_i-\mu)^2}{2\sigma^2}}.$$

The joint p.d.f. for the vector $\mathbf{y}$ is the product of the p.d.f.s for the individual random variables, because the model specifies independence of the y_i, i.e.

$$f(\mathbf{y}) = \prod_{i=1}^{60} f(y_i).$$

2. The New Haven temperature data seem to be 'heavy tailed' relative to a normal: that is, there are more extreme values than are implied by a normal with the observed standard deviation. A better model might be

$$\frac{y_i - \mu}{\sigma} \sim t_\alpha,$$

where μ, σ and α are unknown parameters. Denoting the p.d.f. of a t_α distribution as f_{t_α}, the transformation theory of Section 1.7, combined with independence of the y_i, implies that the p.d.f. of $\mathbf{y}$ is

$$f(\mathbf{y}) = \prod_{i=1}^{60} \frac{1}{\sigma} f_{t_\alpha}\{(y_i - \mu)/\sigma\}.$$

3. Air temperature, a_i, is measured at times t_i (in hours) spaced half an hour apart for a week. The temperature is believed to follow a daily cycle, with a long-term drift over the course of the week, and to be subject to random autocorrelated departures from this overall pattern. A suitable model might then be

$$a_i = \theta_0 + \theta_1 t_i + \theta_2 \sin(2\pi t_i/24) + \theta_3 \cos(2\pi t_i/24) + e_i,$$

where $e_i = \rho e_{i-1} + \epsilon_i$ and the ϵ_i are i.i.d. $N(0, \sigma^2)$. This model implicitly defines the p.d.f. of $\mathbf{a}$, but as specified we have to do a little work to actually find it. Writing $\mu_i = \theta_0 + \theta_1 t_i + \theta_2 \sin(2\pi t_i/24) + \theta_3 \cos(2\pi t_i/24)$, we have $a_i = \mu_i + e_i$. Because e_i is a weighted sum of zero mean normal random variables, it is itself a zero mean normal random variable, with covariance matrix $\boldsymbol{\Sigma}$ such that $\Sigma_{i,j} = \rho^{|i-j|}$. So the p.d.f. of $\mathbf{a}$,[1] the vector of temperatures, must be multivariate normal,

$$f_{\mathbf{a}}(\mathbf{a}) = \frac{1}{\sqrt{(2\pi)^n|\mathbf{\Sigma}|}} e^{-\frac{1}{2}(\mathbf{a}-\boldsymbol{\mu})^{\mathrm{T}}\mathbf{\Sigma}^{-1}(\mathbf{a}-\boldsymbol{\mu})},$$

where $\mathbf{\Sigma}$ depends on parameter ρ and σ, while $\boldsymbol{\mu}$ depends on parameters $\boldsymbol{\theta}$ and covariate $\mathbf{t}$ (see also Section 1.6).

4. Data were collected at the Ohio State University Bone Marrow Transplant Unit to compare two methods of bone marrow transplant for 23 patients suffering from non-Hodgkin's lymphoma. Each patient was randomly allocated to one of two treatments. The *allogenic* treatment consisted of a transplant from a matched sibling donor. The *autogenic* treatment consisted of removing the patient's marrow, 'cleaning it' and returning it after a high dose of chemotherapy. For each patient the time of death, relapse or last follow up (still healthy) is recorded. The 'right-censored' last follow up times are marked with an over-bar.

	Time (Days)										
Allo	28	32	49	84	357	$\overline{933}$	$\overline{1078}$	$\overline{1183}$	$\overline{1560}$	$\overline{2114}$	$\overline{2144}$	
Auto	42	53	57	63	81	140	176	$\overline{210}$	252	$\overline{476}$	524	$\overline{1037}$

The data are from Klein and Moeschberger (2003). A reasonable model is that the death or relapse times are observations of independent random variables having exponential distributions with parameters θ_l and θ_u respectively (mean survival times are $\theta_{u/l}^{-1}$). Medically the interesting question is whether the data are consistent with $\theta_l = \theta_u$.

For the allogenic group, denote the time of death, relapse or censoring by t_i. So we have

$$f_l(t_i) = \begin{cases} \theta_l e^{-\theta_l t_i} & \text{uncensored} \\ \int_{t_i}^{\infty} \theta_l e^{-\theta_l t} dt = e^{-\theta_l t_i} & \text{censored} \end{cases}$$

where f_l is a density for an uncensored t_i (death) or a probability of dying after t_i for a censored observation. A similar model applies for the autogenic sample. For the whole dataset we then have

$$f(\mathbf{t}) = \prod_{i=1}^{11} f_l(t_i) \prod_{i=12}^{23} f_u(t_i).$$

[1] For aesthetic reasons I will use phrases such as 'the p.d.f. of $\mathbf{y}$' to mean 'the p.d.f. of the random vector of which $\mathbf{y}$ is an observation'.

2.2 Random effects and autocorrelation

For the example models in the previous section, it was relatively straightforward to go from the model statement to the implied p.d.f. for the data. Often, this was because we could model the data as observations of independent random variables with known and tractable distributions. Not all datasets are so amenable, however, and we commonly require more complicated descriptions of the stochastic structure in the data. Often we require models with multiple levels of randomness. Such multilayered randomness implies autocorrelation in the data, but we may also need to introduce autocorrelation more directly, as in Example 3 in Section 2.1.

Random variables in a model that are not associated with the independent random variability of single observations,[2] are termed *random effects*. The idea is best understood via concrete examples:

1. A trial to investigate a new blood-pressure reducing drug assigns male patients at random to receive the new drug or one of two alternative standard treatments. Patients' age, a_j, and fat mass, f_j, are recorded at enrolment, and their blood pressure reduction is measured at weekly intervals for 12 weeks. In this setup it is clear that there are two sources of random variability that must be accounted for: the random variability from patient to patient, and the random variability from measurement to measurement made on a single patient. Let y_{ij} represent the ith blood-pressure reduction measurement on the jth patient. A suitable model might then be

$$y_{ij} = \gamma_{k(j)} + \beta_1 a_j + \beta_2 f_j + b_j + \epsilon_{ij}, \quad b_j \sim N(0, \sigma_b^2), \quad \epsilon_{ij} \sim N(0, \sigma^2), \tag{2.1}$$

where $k(j) = 1$, 2 or 3 denotes the treatment to which patient j has been assigned. The γ_k, βs and σs are unknown model parameters. The random variables b_j and ϵ_{ij} are all assumed to be independent here.

The key point is that we decompose the randomness in y_{ij} into two components: (i) the patient specific component, b_j, which varies randomly from patient to patient but remains fixed between measurements on the same patient, and (ii) the individual measurement variability, ϵ_{ij}, which varies between all measurements. Hence measurements taken from different patients of the same age, fat mass and treatment will usually differ more than measurements taken on the same patient. So the y_{ij} are not statistically independent in this model, unless we condition on the b_j.

[2] and, in a Bayesian context, are not parameters.

On first encountering such models it is natural to ask why we do not simply treat the b_j as fixed parameters, in which case we would be back in the convenient world of independent measurements. The reason is interpretability. As stated, (2.1) treats patients as being randomly sampled from a wide population of patients: the patient-specific effects are simply random draws from some normal distribution describing the distribution of patient effects over the patient population. In this setup there is no problem using statistics to make inferences about the population of patients in general, on the basis of the sample of patients in the trial. Now suppose we treat the b_j as parameters. This is equivalent to saying that the patient effects are *entirely* unpredictable from patient to patient – there is no structure to them at all and they could take any value whatsoever. This is a rather extreme position to hold and implies that we can say nothing about the blood pressure of a patient who is not in the trial, because their b_j value could be anything at all. Another side of this problem is that we lose all ability to say anything meaningful about the treatment effects, γ_k, since we have different patients in the different treatment arms, so that the fixed b_j are completely confounded with the γ_k (as can be seen by noting that any constant could be added to a γ_k, while simultaneously being subtracted from all the b_j for patients in group k, without changing the model distribution of any y_{ij}).

2. A population of cells in an experimental chemostat is believed to grow according to the model

$$N_{t+1} = rN_t \exp(-\alpha N_t + b_t), \quad b_t \sim N(0, \sigma_b^2),$$

where N_t is the population at day t; r, α, σ_b and N_0 are parameters; and the b_t are independent random effects. A random sample of 0.5% of the cells in the chemostat is counted every 2 days, giving rise to observations y_t, which can be modelled as independent $\text{Poi}(0.005N_t)$. In this case the random effects enter the model nonlinearly, introducing a complicated correlation structure into N_t, and hence also the y_t.

The first example is an example of a *linear mixed model*.[3] In this case it is not difficult to obtain the p.d.f. for the vector $\mathbf{y}$. We can write the model in matrix vector form as

$$\mathbf{y} = \mathbf{X}\boldsymbol{\beta} + \mathbf{Z}\mathbf{b} + \boldsymbol{\epsilon}, \quad \mathbf{b} \sim N(\mathbf{0}, \mathbf{I}\sigma_b^2), \quad \boldsymbol{\epsilon} \sim N(\mathbf{0}, \mathbf{I}\sigma^2), \tag{2.2}$$

[3] It is a mixed model because it contains both fixed effects (the γ and β terms in the example) and random effects. Mixed models should not be confused with *mixture* models in which each observation is modelled as having some probability of being drawn from each of a number of alternative distributions.

where $\boldsymbol{\beta}^{\mathrm{T}} = (\gamma_1, \gamma_2, \gamma_3, \beta_1, \beta_2)$. The first three columns of $\mathbf{X}$ contain 0, 1 indicator variables depending on which treatment the row relates to, and the next two columns contain the age and fat mass for the patients. $\mathbf{Z}$ has one column for each subject, each row of which contains a 1 or a 0 depending on whether the observation at this data row relates to the subject or not. Given this structure it follows (see Section 1.6.2) that the covariance matrix for $\mathbf{y}$ is $\boldsymbol{\Sigma} = \mathbf{I}\sigma^2 + \mathbf{Z}\mathbf{Z}^{\mathrm{T}}\sigma_b^2$ and the expected value of $\mathbf{y}$ is $\boldsymbol{\mu} = \mathbf{X}\boldsymbol{\beta}$, so that $\mathbf{y} \sim N(\boldsymbol{\mu}, \boldsymbol{\Sigma})$, with p.d.f. as in (1.6). So in this case the p.d.f. for $\mathbf{y}$ is quite easy to write down. However, computing with it can become very costly if the dimension of $\mathbf{y}$ is beyond the low thousands. Hence the main challenge with these models is to find ways of exploiting the sparsity that results from having so many 0 entries in $\mathbf{Z}$, so that computation is feasible for large samples.

The second example illustrates the more usual situation in which the model fully specifies a p.d.f. (or p.f.) for $\mathbf{y}$, but it is not possible to write it down in closed form, or even to evaluate it exactly. In contrast, the joint density of the random effects, $\mathbf{b}$, and data, $\mathbf{y}$, is always straightforward to evaluate. From Sections 1.4.2 and 1.4.3 we have that

$$f(\mathbf{y}, \mathbf{b}) = f(\mathbf{y}|\mathbf{b})f(\mathbf{b}),$$

and the distributions $f(\mathbf{y}|\mathbf{b})$ and $f(\mathbf{b})$ are usually straightforward to work with. So, for the second example, let $f(y; \lambda)$ denote the p.f. of a Poisson random variable with mean λ (see Section A.3.2). Then

$$f(\mathbf{y}|\mathbf{b}) = \prod_t f(y_t; N_t/200),$$

while $f(\mathbf{b})$ is the density of a vector of i.i.d. $N(0, \sigma_b^2)$ deviates.

For some statistical tasks we may be able to work directly with $f(\mathbf{y}, \mathbf{b})$ without needing to evaluate the p.d.f. of $\mathbf{y}$: this typically applies when taking the Bayesian approach of Section 2.5, for example. However, often we cannot escape the need to evaluate $f(\mathbf{y})$ itself. That is, we need

$$f(\mathbf{y}) = \int \mathbf{f}(\mathbf{y}, \mathbf{b})d\mathbf{b},$$

which is generally not analytically tractable. We then have a number of choices. If the model has a structure that allows the integral to be broken down into a product of low-dimensional integrals then numerical integration methods (so-called quadrature) may be feasible; however, these methods are usually impractical beyond somewhere around 10 dimensions. Then we need a different approach: either estimate the integral statistically using stochastic simulation or approximate it somehow (see Section 5.3.1).

2.3 Inferential questions

Given some data, $\mathbf{y}$, and a statistical model with parameters $\boldsymbol{\theta}$, there are four basic questions to ask:

1. What values for $\boldsymbol{\theta}$ are most consistent with $\mathbf{y}$?
2. Is some prespecified restriction on $\boldsymbol{\theta}$ consistent with $\mathbf{y}$?
3. What ranges of values of $\boldsymbol{\theta}$ are consistent with $\mathbf{y}$?
4. Is the model consistent with the data for any values of $\boldsymbol{\theta}$ at all?

The answers to these questions are provided by *point estimation*, *hypothesis testing*, *interval estimation* and *model checking*, respectively. Question 2 can be somewhat generalised to: which of several alternative models is most consistent with y? This is the question of *model selection* (which partly incorporates question 4). Central to the statistical way of doing things is recognising the uncertainty inherent in trying to learn about $\boldsymbol{\theta}$ from $\mathbf{y}$. This leads to another, often neglected, question that applies when there is some control over the data-gathering process:

5. How might the data-gathering process be organized to produce data that enables answers to the preceding questions to be as accurate and precise as possible?

This question is answered by experimental and survey *design* methods.

There are two main classes of methods for answering questions 1-4, and they start from different basic assumptions. These are the *Bayesian* and *frequentist* approaches, which differ in how they use probability to model uncertainty about model parameters. In the frequentist approach, parameters are treated as having values that are fixed states of nature, about which we want to learn using data. There is randomness in our estimation of the parameters, but not in the parameters themselves. In the Bayesian approach parameters are treated as random variables, about which we want to update our beliefs in the light of data: our beliefs are summarised by probability distributions for the parameters. The difference between the approaches can sound huge, and there has been much debate about which is least ugly. From a practical perspective, however, the approaches have much in common, except perhaps when it comes to model selection. In particular, if properly applied they usually produce results that differ by less than the analysed models are likely to differ from reality.

2.4 The frequentist approach

In this way of doing things we view parameters, $\boldsymbol{\theta}$, as fixed states of nature, about which we want to learn. We use probability to investigate what would happen under repeated replication of the data (and consequent statistical analysis). In this approach probability is all about how *frequently* events would occur under this imaginary replication process.

2.4.1 Point estimation: maximum likelihood

Given a model and some data, then with enough thought about what the unknown model parameters mean, it is often possible to come up with a way of getting reasonable parameter value guesses from the data. If this process can be written down as a mathematical recipe, then we can call the guess an estimate, and we can study its properties under data replication to get an idea of its uncertainty. But such model-by-model reasoning is time consuming and somewhat unsatisfactory: how do we know that our estimation process is making good use of the data, for example? A general approach for dealing with all models would be appealing.

There are a number of more or less general approaches, such as the *method of moments* and *least squares* methods, which apply to quite wide classes of models, but one general approach stands out in terms of practical utility and nice theoretical properties: *maximum likelihood estimation*. The key idea is simply this:

Parameter values that make the observed data appear relatively probable are more *likely* to be correct than parameter values that make the observed data appear relatively improbable.

For example, we would much prefer an estimate of $\boldsymbol{\theta}$ that assigned a probability density of 0.1 to our observed $\mathbf{y}$, according to the model, to an estimate for which the density was 0.00001.

So the idea is to judge the *likelihood* of parameter values using $f_\theta(\mathbf{y})$, the model p.d.f. according to the given value of $\boldsymbol{\theta}$, evaluated at the observed data. Because $\mathbf{y}$ is now fixed and we are considering the likelihood as a function of $\boldsymbol{\theta}$, it is usual to write the likelihood as $L(\boldsymbol{\theta}) \equiv f_\theta(\mathbf{y})$. In fact, for theoretical and practical purposes it is usual to work with the log likelihood $l(\boldsymbol{\theta}) = \log L(\boldsymbol{\theta})$. The *maximum likelihood estimate* (MLE) of $\boldsymbol{\theta}$ is then

$$\hat{\boldsymbol{\theta}} = \underset{\theta}{\operatorname{argmax}}\, l(\boldsymbol{\theta}).$$

There is more to maximum likelihood estimation than just its intuitive appeal. To see this we need to consider what might make a good estimate, and to do that we need to consider repeated estimation under repeated replication of the data-generating process.

Replicating the random data and repeating the estimation process results in a different value of $\hat{\boldsymbol{\theta}}$ for each replicate. These values are of course observations of a random vector, the *estimator* or $\boldsymbol{\theta}$, which is usually also denoted $\hat{\boldsymbol{\theta}}$ (the context making clear whether estimate or estimator is being referred to). Two theoretical properties are desirable:

1. $E(\hat{\boldsymbol{\theta}}) = \boldsymbol{\theta}$ or at least $|E(\hat{\boldsymbol{\theta}}) - \boldsymbol{\theta}|$ should be small (i.e. the estimator should be *unbiased*, or have small bias).
2. $\text{var}(\hat{\boldsymbol{\theta}})$ should be small (i.e. the estimator should have low variance).

Unbiasedness basically says that the estimator gets it right on average: a long-run average of the $\hat{\boldsymbol{\theta}}$, over many replicates of the data set, would tend towards the true value of the parameter vector. Low variance implies that any individual estimate is quite precise. There is a tradeoff between the two properties, so it is usual to seek both. For example, we can always drive variance to zero if we do not care about bias, by just eliminating the data from the estimation process and picking a constant for the estimate. Similarly it is easy to come up with all sorts of unbiased estimators that have enormous variance. Given the tradeoff, you might reasonably wonder why we do not concern ourselves with some direct measure of estimation error such as $E\{(\hat{\boldsymbol{\theta}}-\boldsymbol{\theta})^2\}$, the mean square error (MSE). The reason is that it is difficult to prove general results about minimum MSE estimators, so we are stuck with the second-best option of considering minimum variance unbiased estimators.[4]

It is possible to derive a lower limit on the variance that any unbiased estimator can achieve: the Cramér-Rao lower bound. Under some regularity conditions, and in the large sample limit, it turns out that maximum likelihood estimation is unbiased and achieves the Cramér-Rao lower bound, which gives some support for its use (see Sections 4.1 and 4.3). In addition, under the same conditions,

$$\hat{\boldsymbol{\theta}} \sim N(\boldsymbol{\theta}, \boldsymbol{\mathcal{I}}^{-1}), \tag{2.3}$$

[4] Unless the gods have condemned you to repeat the same experiment for all eternity, unbiasedness, although theoretically expedient, should not be of much intrinsic interest: an estimate close to the truth for the data at hand should always be preferable to one that would merely get things right on average over an infinite sequence of data replicates.

where $\mathcal{I}_{ij} = -E(\partial^2 l/\partial\theta_i\partial\theta_j)$ (and actually the same result holds substituting $\hat{\mathcal{I}}_{ij} = -\partial^2 l/\partial\theta_i\partial\theta_j$ for $\mathcal{I}_{ij}$).

2.4.2 Hypothesis testing and p-values

Now consider the question of whether some defined restriction on $\boldsymbol{\theta}$ is consistent with $\mathbf{y}$?

p-values: the fundamental idea

Suppose that we have a model defining a p.d.f., $f_{\boldsymbol{\theta}}(\mathbf{y})$, for data vector $\mathbf{y}$ and that we want to test the *null hypothesis*, $\text{H}_0 : \boldsymbol{\theta} = \boldsymbol{\theta}_0$, where $\boldsymbol{\theta}_0$ is some specified value. That is, we want to establish whether the data could reasonably be generated from $f_{\theta_0}(\mathbf{y})$. An obvious approach is to ask how probable data like $\mathbf{y}$ are under H_0. It is tempting to simply evaluate $f_{\theta_0}(\mathbf{y})$ for the observed $\mathbf{y}$, but then deciding what counts as 'probable' and 'improbable' is difficult to do in a generally applicable way.

A better approach is to assess the probability, p_0 say, of obtaining data at least as improbable as $\mathbf{y}$ under H_0 (better read that sentence twice). For example, if only one dataset in a million would be as improbable as $\mathbf{y}$, according to H_0, then assuming we believe our data, we ought to seriously doubt H_0. Conversely, if half of all datasets would be expected to be at least as improbable as $\mathbf{y}$, according to H_0, then there is no reason to doubt it.

A quantity like p_0 makes good sense in the context of goodness of fit testing, where we simply want to assess the plausibility of f_{θ_0} as a model without viewing it as being a restricted form of a larger model. But when we are really testing $\text{H}_0 : \boldsymbol{\theta} = \boldsymbol{\theta}_0$ against the alternative H_1 : '$\boldsymbol{\theta}$ unrestricted' then p_0 is not satisfactory, because it makes no distinction between $\mathbf{y}$ being improbable under H_0 but probable under H_1, and $\mathbf{y}$ being improbable under both.

A very simple example illustrates the problem. Consider independent observations y_1, y_2 from $N(\mu, 1)$, and the test $\text{H}_0 : \mu = 0$ versus $\text{H}_0 : \mu \neq 0$. Figure 2.2 shows the p.d.f. under the null, and, in grey, the region over which the p.d.f. has to be integrated to find p_0 for the data point marked by •. Now consider two alternative values for y_1, y_2 that yield equal $p_0 = 0.1$. In one case (black triangle) $y_1 = -y_2$, so that the best estimate of μ is 0, corresponding exactly to H_0. In the other case (black circle) the data are much more probable under the alternative than under the null hypothesis. So because we include points that are more compatible with the null than with the alternative in the calculation of p_0, we have only weak discriminatory power between the hypotheses.

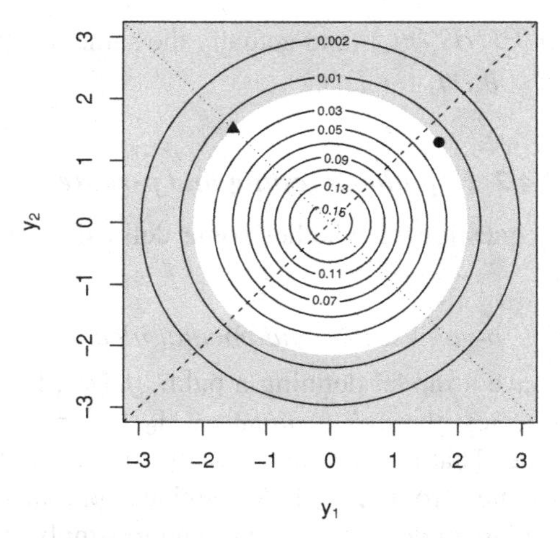

Figure 2.2 Why the alternative hypothesis is needed in defining the p-value, and p_0 will not do. Contour plot of the joint p.d.f. for the null model that y_1, y_2 are independent $N(0, 1)$. The dashed line illustrates the possible values for the expected values of y_1 and y_2, under the alternative model that they are independent $N(\mu, 1)$. The black triangle and black circle show two possible values for y_1, y_2, while the grey region shows the region of at least as improbable y_1, y_2 pairs, corresponding to $p_0 = 0.1$. The problem is that although the black circle is much more probable under the alternative model, it has the same p_0 value as the black triangle, for which $y_1 = -y_2$ and the estimated μ would be exactly the null model value of zero. The dotted line is $y_1 = -y_2$.

Recognizing the problems with p_0, a possible solution is to standardise $f_{\theta_0}(\mathbf{y})$ by the highest value that $f_\theta(\mathbf{y})$ could have taken for the given $\mathbf{y}$. That is, to judge the *relative plausibility* of $\mathbf{y}$ under H_0 on the basis of $f_{\theta_0}(\mathbf{y})/f_{\hat{\theta}}(\mathbf{y})$ where $\hat{\boldsymbol{\theta}}$ is the value maximising $f_\theta(\mathbf{y})$ for a given $\mathbf{y}$. In the context of the example in Figure 2.2 this approach is much better. The black triangle now has relative plausibility 1, reflecting its compatibility with the H_0, whereas the black circle has much lower plausibility, reflecting the fact that it would be much more probable under a model with a mean greater than zero. So we could now seek a revised measure of consistency of the data and null hypothesis:

p is the probability, under the null hypothesis, of obtaining data at least as relatively implausible as that observed.

Actually the reciprocal of this relative plausibility is generally known as the *likelihood ratio* $f_{\hat{\theta}}(\mathbf{y})/f_{\theta_0}(\mathbf{y})$ of the two hypotheses, because it is a measure of how likely the alternative hypothesis is relative to the null, given the data. So we have the more usual equivalent definition:

p is the probability, under the null hypothesis, of obtaining a likelihood ratio at least as large as that observed.

p is generally referred to as the *p-value* associated with a test. If the null hypothesis is true, then from its definition, the p-value should have a uniform distribution on $[0, 1]$ (assuming its distribution is continuous). By convention p-values in the ranges $0.1 \geq p > 0.05$, $0.05 \geq p > 0.01$, $0.01 \geq p > 0.001$ and $p \leq 0.001$ are sometimes described as providing, respectively, 'marginal evidence', 'evidence', 'strong evidence' and 'very strong evidence' against the null model, although the interpretation should really be sensitive to the context.

Generalisations

For the purposes of motivating p-values, the previous subsection considered only the case where the null hypothesis is a *simple* hypothesis, specifying a value for every parameter of f, while the alternative is a *composite* hypothesis, in which a range of parameter values are consistent with the alternative. Unsurprisingly, there are many situations in which we are interested in comparing two composite hypotheses, so that H_0 specifies some restrictions of $\boldsymbol{\theta}$, without fully constraining it to one point. Less commonly, we may also wish to compare two simple hypotheses, so that the alternative also supplies one value for each element of $\boldsymbol{\theta}$. This latter case is of theoretical interest, but because the hypotheses are not nested it is somewhat conceptually different from most cases of interest.

All test variants can be dealt with by a slight generalisation of the likelihood ratio statistic to $f_{\hat{\theta}}(\mathbf{y})/f_{\hat{\theta}_0}(\mathbf{y})$ where $f_{\hat{\theta}_0}(\mathbf{y})$ now denotes the maximum possible value for the density of $\mathbf{y}$ under the null hypothesis. If the null hypothesis is simple, then this is just $f_{\theta_0}(\mathbf{y})$, as before, but if not then it is obtained by finding the parameter vector that maximises $f_{\theta}(\mathbf{y})$ subject to the restrictions on $\boldsymbol{\theta}$ imposed by H_0.

In some cases the p-value can be calculated exactly from its definition, and the relevant likelihood ratio. When this is not possible, there is a large sample result that applies in the usual case of a composite alternative with a simple or composite null hypothesis. In general we test $H_0 : \mathbf{R}(\boldsymbol{\theta}) = \mathbf{0}$ against H_1 : '$\boldsymbol{\theta}$ unrestricted', where $\mathbf{R}$ is a vector-valued function of $\boldsymbol{\theta}$, specifying r restrictions on $\boldsymbol{\theta}$. Given some regularity conditions and in the large sample limit,

$$2\{\log f_{\hat{\theta}}(\mathbf{y}) - \log f_{\hat{\theta}_0}(\mathbf{y})\} \sim \chi^2_r, \tag{2.4}$$

under H_0. See Section 4.4.

$f_{\hat{\theta}}(\mathbf{y})/f_{\hat{\theta}_0}(\mathbf{y})$ is an example of a *test statistic*, which takes low values when the H_0 is true, and higher values when H_1 is true. Other test statistics can be devised in which case the definition of the p-value generalises to:

p is the probability of obtaining a test statistic at least as favourable to H_1 as that observed, if H_0 is true.

This generalisation immediately raises the question: what makes a good test statistic? The answer is that we would like the resulting p-values to be as small as possible when the null hypothesis is not true (for a test statistic with a continuous distribution, the p-values should have a $U(0, 1)$ distribution when the null is true). That is, we would like the test statistic to have *high power* to discriminate between null and alternative hypotheses.

The Neyman-Pearson lemma

The Neyman-Pearson lemma provides some support for using the likelihood ratio as a test statistic, in that it shows that doing so provides the best chance of rejecting a false null hypothesis, albeit in the restricted context of a simple null versus a simple alternative. Formally, consider testing $H_0 : \boldsymbol{\theta} = \boldsymbol{\theta}_0$ against $H_1 : \boldsymbol{\theta} = \boldsymbol{\theta}_1$. Suppose that we decide to reject H_0 if the p-value is less than or equal to some value α. Let $\beta(\boldsymbol{\theta})$ be the probability of rejection if the true parameter value is $\boldsymbol{\theta}$ – the test's *power*.

In this accept/reject setup the likelihood ratio test rejects H_0 if $\mathbf{y} \in R = \{\mathbf{y} : f_{\theta_1}(\mathbf{y})/f_{\theta_0}(\mathbf{y}) > k\}$ and k is such that $\Pr_{\theta_0}(\mathbf{y} \in R) = \alpha$. It is useful to define the function $\phi(\mathbf{y}) = 1$ if $\mathbf{y} \in R$ and 0 otherwise. Then $\beta(\boldsymbol{\theta}) = \int \phi(\mathbf{y}) f_{\theta}(\mathbf{y}) d\mathbf{y}$. Note that $\beta(\boldsymbol{\theta}_0) = \alpha$.

Now consider using an alternative test statistic and again rejecting if the p-value is $\leq \alpha$. Suppose that the test procedure rejects if

$$\mathbf{y} \in R^* \text{ where } \Pr_{\theta_0}(\mathbf{y} \in R^*) \leq \alpha.$$

Let $\phi^*(\mathbf{y})$ and $\beta^*(\boldsymbol{\theta})$ be the equivalent of $\phi(\mathbf{y})$ and $\beta(\boldsymbol{\theta})$ for this test. Here $\beta^*(\boldsymbol{\theta}_0) = \Pr_{\theta_0}(\mathbf{y} \in R^*) \leq \alpha$.

The *Neyman-Pearson Lemma* then states that $\beta(\boldsymbol{\theta}_1) \geq \beta^*(\boldsymbol{\theta}_1)$ (i.e. the likelihood ratio test is the most powerful test possible).

Proof follows from the fact that

$$\{\phi(\mathbf{y}) - \phi^*(\mathbf{y})\}\{f_{\theta_1}(\mathbf{y}) - kf_{\theta_0}(\mathbf{y})\} \geq 0,$$

since from the definition of R, the first bracket is non-negative whenever the second bracket is non-negative, and it is non-positive whenever the second bracket is negative. In consequence,

$$\begin{aligned} 0 &\leq \int \{\phi(\mathbf{y}) - \phi^*(\mathbf{y})\}\{f_{\theta_1}(\mathbf{y}) - kf_{\theta_0}(\mathbf{y})\}d\mathbf{y} \\ &= \beta(\boldsymbol{\theta}_1) - \beta^*(\boldsymbol{\theta}_1) - k\{\beta(\boldsymbol{\theta}_0) - \beta^*(\boldsymbol{\theta}_0)\} \leq \beta(\boldsymbol{\theta}_1) - \beta^*(\boldsymbol{\theta}_1), \end{aligned}$$

since $\{\beta(\boldsymbol{\theta}_0) - \beta^*(\boldsymbol{\theta}_0)\} \geq 0$. So the result is proven. Casella and Berger (1990) give a fuller version of the lemma, on which this proof is based.

2.4.3 Interval estimation

Recall the question of finding the range of values for the parameters that are consistent with the data. An obvious answer is provided by the range of values for any parameter θ_i that would have been accepted in a hypothesis test. For example, we could look for all values of θ_i that would have resulted in a p-value of more than 5% if used as a null hypothesis for the parameter. Such a set is known as a 95% *confidence set* for θ_i. If the set is continuous then its endpoints define a 95% *confidence interval.*

The terminology comes about as follows. Recall that if we reject a hypothesis when the p-values is less than 5% then we will reject the null on 5% of occasions when it is correct and therefore accept it on 95% of occasions when it is correct. This follows directly from the definition of a p-value and the fact that it has a $U(0, 1)$ distribution when the null hypothesis is correct.[5] Clearly if the test rejects the true parameter value 5% of the time, then the corresponding confidence intervals must exclude the true parameter value on those 5% of occasions as well. That is, a 95% confidence interval has a 0.95 probability of including the true parameter value (where the probability is taken over an infinite sequence of replicates of the data-gathering and intervals estimation process). The following graphic shows 95% confidence intervals computed from 20 replicate datasets, for a single parameter θ with true value θ_{true}.

[5] again assuming a continuously distributed test statistic. In the less common case of a discretely distributed test statistic, then the distribution will not be exactly $U(0, 1)$.

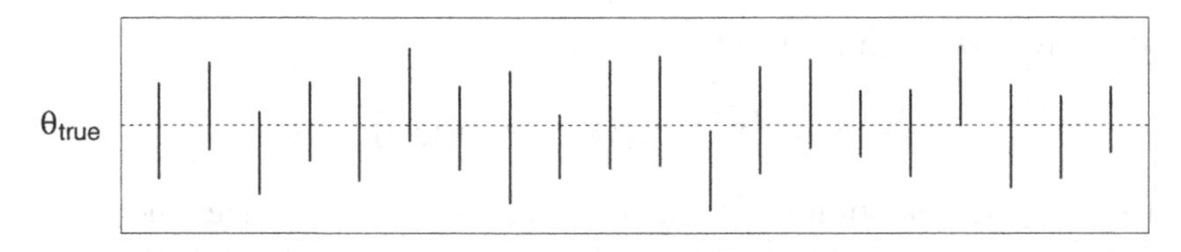

As expected on average, 19 of the intervals include the true value, and 1 does not. In general,

A $\gamma 100\%$ confidence interval for θ is (an observation of) a random interval designed to have a probability γ of including the true value of θ.

Again, maximum likelihood theory provides general recipes for computing intervals that will be correct in the large sample limit. We can either base intervals on result (2.3) and Section 2.7 or search for the range of θ_i values giving p-values above $1 - \gamma$, in a test using (2.4). The latter *profile likelihood* intervals have the advantage that parameters inside the interval have higher likelihood than those outside it.

2.4.4 Model checking

Ultimately a statistical model says that our data, $\mathbf{y}$, are observations of a random vector with probability density function $f_\theta(\mathbf{y})$. That is, the model says that $\mathbf{y} \sim f_\theta(\mathbf{y})$. The aim of model checking is to show that

$$\mathbf{y} \not\sim f_\theta(\mathbf{y}),$$

i.e. to show that the model is wrong in some serious and detectable way.

In most cases we know that the model is wrong: it is a model, not reality. The point is to look for ways in which the model is so wrong that any conclusions we might want to draw from it become questionable. The idea is that if we cannot detect that the model is wrong statistically, then statistical conclusions drawn with its aid are likely to be reasonably reliable.[6]

No single test or informal check can detect all possible ways in which a model might be wrong. Model checking calls for judgement and 'quantitative scepticism'. Often the most useful checks are graphical ones, because when they indicate that a model is wrong, they frequently also indicate *how* it is wrong. One plot that can be produced for any model is a quantile-quantile (QQ) plot of the marginal distribution of the elements of $\mathbf{y}$, in which the sorted elements of $\mathbf{y}$ are plotted against quantiles of the model

[6] More cautiously, if we can statistically detect that the model is wrong, then statistical conclusions drawn from it are very likely to be wrong.

distribution of $\mathbf{y}$. Even if the quantile function is not tractable, replicate $\mathbf{y}$ vectors can be repeatedly simulated from $f_{\hat{\theta}}(\mathbf{y})$, and we can obtain empirical quantiles for the marginal distribution of the simulated y_i. An approximately straight line plot should result if all is well (and reference bands for the plot can also be obtained from the simulations).

But such marginal plots will not detect all model problems, and more is usually needed. Often a useful approach is to examine plots of standardised *residuals*. The idea is to remove the modelled systematic component of the data and to look at what is left over, which should be random. Typically the residuals are standardised so that if the model is correct they should appear independent with constant variance. Exactly how to construct useful residuals is model dependent, but one fairly general approach is as follows. Suppose that the fitted model implies that the expected value and covariance matrix of $\mathbf{y}$ are $\boldsymbol{\mu}_{\hat{\theta}}$ and $\boldsymbol{\Sigma}_{\hat{\theta}}$. Then we can define standardised residuals

$$\hat{\epsilon} = \boldsymbol{\Sigma}_{\hat{\theta}}^{-1/2}(\mathbf{y} - \boldsymbol{\mu}_{\hat{\theta}}),$$

which should appear to be approximately independent, with zero mean and unit variance, if the model is correct. $\boldsymbol{\Sigma}_{\hat{\theta}}^{-1/2}$ is any matrix square root of $\boldsymbol{\Sigma}_{\hat{\theta}}^{-1}$, for example its Choleski factor (see Appendix B). Of course, if the elements of $\mathbf{y}$ are independent according to the model, then the covariance matrix is diagonal, and the computations are very simple.

The standardised residuals are then plotted against $\boldsymbol{\mu}_{\hat{\theta}}$, to look for patterns in their mean or variance, which would indicate something missing in the model structure or something wrong with the distributional assumption, respectively. The residuals would also be plotted against any covariates in the model, with similar intention. When the data have a temporal element then the residuals would also be examined for correlations in time. The basic idea is to try to produce plots that show in some way that the residuals are not independent with constant/unit variance. Failure to find such plots increases faith in the model.

2.4.5 Further model comparison, AIC and cross-validation

One way to view the hypothesis tests of Section 2.4.2 is as the comparison of two alternative models, where the null model is a simplified (restricted) version of the alternative model (i.e where the models are *nested*). The methods of Section 2.4.2 are limited in two major respects. First, they provide no general way of comparing models that are not nested, and second, they are based on the notion that we want to stick with the null model until there is strong evidence to reject it. There is an obvious need for model comparison methods that simply seek the 'best' model, from some set of models that need not necessarily be nested.

Akaike's information criterion (AIC; Akaike, 1973) is one attempt to fill this need. First we need to formalise what 'best' means in this context: closest to the underlying true model seems sensible. We saw in Section 2.4.2 that the likelihood ratio, or its log, is a good way to discriminate between models, so a good way to measure model closeness might be to use the expected value of the log likelihood ratio of the true model and the model under consideration:

$$K(f_{\hat{\theta}}, f_t) = \int \{\log f_t(\mathbf{y}) - \log f_{\hat{\theta}}(\mathbf{y})\} f_t(\mathbf{y}) d\mathbf{y}$$

where f_t is the true p.d.f. of $\mathbf{y}$. K is known as the *Kullback-Leibler* divergence (or distance). Selecting models to minimise an estimate of the expected value of K (expectation over the distribution of $\hat{\boldsymbol{\theta}}$) is equivalent to selecting the model that has the lowest value of

$$\text{AIC} = -2l(\hat{\boldsymbol{\theta}}) + 2\text{dim}(\boldsymbol{\theta}).$$

See Section 4.6 for a derivation.

Notice that if we were to select models only on the basis of which has the highest likelihood, we would encounter a fundamental problem: even if a parameter is not in the true model, the extra flexibility it provides means that adding it never decreases the maximised likelihood and almost always increases it. So likelihood almost always selects the more complex model. AIC overcomes this problem by effectively adding a penalty for adding parameters: if a parameter is not needed, the AIC is unlikely to decrease when it is added to the model.

An alternative recognises that the KL divergence only depends on the model via $-\int \log f_{\hat{\theta}}(\mathbf{y}) f_t(\mathbf{y}) d\mathbf{y}$, the expectation of the model maximised log likelihood, where the expectation is taken over data not used to estimate $\hat{\boldsymbol{\theta}}$. An obvious direct estimator of this is the *cross-validation* score

$$\text{CV} = -\sum_i \log f_{\hat{\theta}^{[-i]}}(y_i),$$

where $\hat{\theta}^{[-i]}$ is the MLE based on the data with y_i omitted (i.e. we measure the average ability of the model to predict data to which it was not fitted). Sometimes this can be computed or approximated efficiently, and variants are possible in which more than one data point at a time are omitted from fitting. However, in general it is more costly than AIC.

2.5 The Bayesian approach

The other approach to answering the questions posed in Section 2.3 is the Bayesian approach. This starts from the idea that $\boldsymbol{\theta}$ is itself a random vector and that we can describe our prior knowledge about $\boldsymbol{\theta}$ using a *prior* probability distribution. The main task of statistical inference is then to update our knowledge (or at any rate beliefs) about $\boldsymbol{\theta}$ in the light of data $\mathbf{y}$. Given that the parameters are now random variables, it is usual to denote the model likelihood as the conditional distribution $f(\mathbf{y}|\boldsymbol{\theta})$. Our updated beliefs about $\boldsymbol{\theta}$ are then expressed using the *posterior* density

$$f(\boldsymbol{\theta}|\mathbf{y}) = \frac{f(\mathbf{y}|\boldsymbol{\theta})f(\boldsymbol{\theta})}{f(\mathbf{y})}, \tag{2.5}$$

which is just Bayes theorem from Section 1.4.3 (again f with different arguments are all different functions here). The likelihood, $f(\mathbf{y}|\boldsymbol{\theta})$, is specified by our model, exactly as before, but the need to specify the prior, $f(\boldsymbol{\theta})$, is new. Note one important fact: it is often not necessary to specify a proper distribution for $f(\boldsymbol{\theta})$ in order for $f(\mathbf{y}|\boldsymbol{\theta})$ to be proper. This opens up the possibility of using improper uniform priors for $\boldsymbol{\theta}$; that is, specifying that $\boldsymbol{\theta}$ can take any value with equal probability density.[7]

Exact computation of (2.5) is rarely possible for interesting models, but it is possible to simulate from $f(\boldsymbol{\theta}|\mathbf{y})$ and often to approximate it, as we see later. For the moment we are interested in how the inferential questions are answered under this framework.

2.5.1 Posterior modes

Under the Bayesian paradigm we do not estimate parameters: rather we compute a whole distribution for the parameters given the data. Even so, we can still pose the question of which parameters are most consistent with the data. A reasonable answer is that it is the most probable value of $\boldsymbol{\theta}$ according to the posterior: the *posterior mode*,

$$\hat{\boldsymbol{\theta}} = \underset{\theta}{\text{argmax}}\, f(\boldsymbol{\theta}|\mathbf{y}).$$

More formally, we might specify a loss function quantifying the loss associated with a particular $\hat{\boldsymbol{\theta}}$ and use the minimiser of this over the posterior distribution as the estimate. If we specify an improper uniform prior

[7] This is not the same as providing *no* prior information about $\boldsymbol{\theta}$. e.g. assuming that θ has an improper uniform prior distribution is different from assuming the same for $\log(\theta)$.

$f(\boldsymbol{\theta}) = k$, then $f(\boldsymbol{\theta}|\mathbf{y}) \propto f(\mathbf{y}|\boldsymbol{\theta})$ and the posterior modes are exactly the maximum likelihood estimates (given that $f(\mathbf{y})$ does not depend on $\boldsymbol{\theta}$). In fact, for data that are informative about a fixed dimension parameter vector $\boldsymbol{\theta}$, then as the sample size tends to infinity the posterior modes tend to the maximum likelihood estimates in any case, because the prior is then dominated by the likelihood.

2.5.2 Model comparison, Bayes factors, prior sensitivity, BIC, DIC

Hypothesis testing, in the sense of Section 2.4.2, does not fit easily with the Bayesian approach, and a criterion somehow similar to AIC is also not straightforward. The obvious approach to Bayesian model selection is to include all possible models in the analysis and then to compute the marginal posterior probability for each model (e.g. Green, 1995). This sounds clean, but it turns out that those probabilities are sensitive to the priors put on the model parameters, which is problematic when these are 'priors of convenience' rather than well-founded representations of prior knowledge. Computing such probabilities is also not easy. This section examines the issue of sensitivity to priors and then covers two of the attempts to come up with a Bayesian equivalent to AIC. See Section 6.6.4 for an alternative approach based on posterior simulation.

Marginal likelihood, the Bayes factor and sensitivity to priors

In the Bayesian framework the goal of summarising the evidence for or against two alternative models can be achieved by the *Bayes factor* (which therefore plays a somewhat similar role to frequentist p-values). A natural way to compare two models, M_1 and M_0, is via the ratio of their probabilities.[8] As a consequence of Bayes theorem, the prior probability ratio transforms to the posterior probability ratio as

$$\frac{\Pr(M_1|\mathbf{y})}{\Pr(M_0|\mathbf{y})} = \frac{f(\mathbf{y}|M_1)\Pr(M_1)}{f(\mathbf{y}|M_0)\Pr(M_0)} = B_{10}\frac{\Pr(M_1)}{\Pr(M_0)},$$

by definition of B_{10}, which is known as the Bayes factor for comparing M_1 with M_0. So B_{10} measures the amount by which the data have shifted the prior probability ratio in favour of M_1. As with p-values, there are conventions about how to describe the degree of evidence that different

[8] If M_1 and M_0 are the only two possibilities then the probability ratio is also the *odds* of M_1; that is, the probability of M_1 over the probability of not M_1.

magnitudes of Bayes factor represent (Kass and Raftery, 1995). Working with $2 \log B_{10}$ (for comparability with the log likelihood ratio) we have

$2 \log B_{10}$	Evidence against M_0
$0 - 2$	Barely worth mentioning
$2 - 6$	Positive
$6 - 10$	Strong
> 10	Very strong

To actually compute B_{10} we need to obtain the marginal density of $\mathbf{y}$ given each model, also known as the *marginal likelihood*. For example,

$$f(\mathbf{y}|M_1) = \int f(\mathbf{y}|\boldsymbol{\theta}_1) f(\boldsymbol{\theta}_1) d\boldsymbol{\theta}_1, \tag{2.6}$$

where $\boldsymbol{\theta}_1$ denotes the parameters of M_1. The need to integrate over all possible parameter values is a major difference between the Bayes factor and the likelihood ratio statistic for model comparison, but integration gives the Bayes factor some advantage. The likelihood ratio statistic is a ratio of maximised likelihoods evaluated at both models' best fit parameters, giving the larger model an inevitable advantage, which we then have to allow for in interpreting the ratio; hence the need for p-values or AIC. By integrating over all possible parameter values, the marginal likelihood does not suffer from this bias towards large models – irrelevant flexibility can decrease the marginal likelihood. Computing (2.6) is generally not straightforward, with two main lines of attack being via stochastic simulation (see Section 6.3.1), or Laplace approximation of the integral. However, there is also a more fundamental problem to consider.

Examination of (2.6) indicates an immediate problem with the use of vague, uninformative or improper priors (i.e. with any prior that is chosen to represent a broad statement of ignorance, rather than a precise characterisation of prior knowledge). The difficulty is that the value of (2.6) is very sensitive to the prior. It is easy to see the problem by example. Suppose the likelihood indicates that a single parameter θ is almost certain to lie in the interval $(0, 1)$, but because we had no real prior information on θ, we used a $U(-100, 100)$ prior, obtaining a value for the marginal likelihood of k. Now suppose that we replace the prior with $U(-1000, 1000)$. This produces a negligible change in the posterior for θ, but reduces the marginal likelihood to approximately $k/10$ (corresponding to a 'positive' change in the Bayes factor in the earlier table). If we had used an improper prior then it would only have been possible to compute the marginal likelihood to within an arbitrary multiplicative constant, rendering the Bayes factor completely useless unless the same improper priors apply to both models.

Another way of seeing the problem with the marginal likelihood is to recognise that it is the likelihood for the fixed parameters of the prior (that is, for those parameter values we chose during model specification), all other parameters having been integrated out. Choosing between models on the basis of the relative likelihood of the fixed parameters of the prior is not always a natural approach. Indeed it is completely arbitrary if those values were selected merely to be as uninformative as possible about $\boldsymbol{\theta}$.

In summary, because the prior is inescapably part of the model in the Bayesian approach, marginal likelihoods, Bayes factors and posterior model probabilities are inescapably sensitive to the choice of prior. In consequence, it is only when those priors that differ between alternative models are really precise and meaningful representations of prior knowledge that we can justify using Bayes factors and posterior model probabilities for model selection. Even then, the computation of the marginal likelihood is often difficult. These difficulties are part of the motivation for attempting to produce AIC-like model selection criteria (see Section 2.4.5) in the Bayesian setting. But before looking at these, let us consider fixing the Bayes Factor.

Intrinsic, fractional and partial Bayes factors

Given that Bayes factors require meaningfully informative priors, which are often not available at the model formulation stage, it is worth considering the alternative of using part of the data to generate priors. The basic idea is to split the data $\mathbf{y}$ into two parts $\mathbf{x}$ and $\mathbf{z}$, and to use $f(\boldsymbol{\theta}|\mathbf{x})$ as the prior for computing the marginal likelihood based on $\mathbf{z}$. That is, marginal likelihoods of the form

$$f(\mathbf{z}|M_i, \mathbf{x}) = \int f(\mathbf{z}|\boldsymbol{\theta}_i, \mathbf{x}) f(\boldsymbol{\theta}_i|\mathbf{x}) d\boldsymbol{\theta}_i$$

are used to form a sort of partial Bayes factor. To see why this improves matters, substitute $f(\boldsymbol{\theta}_i|\mathbf{x}) = f(\mathbf{x}|\boldsymbol{\theta}_i)f(\boldsymbol{\theta}_i)/\int f(\mathbf{x}|\boldsymbol{\theta}_i)f(\boldsymbol{\theta}_i)d\boldsymbol{\theta}_i$ into the preceding expression to get

$$\begin{aligned} f(\mathbf{z}|M_i, \mathbf{x}) &= \frac{\int f(\mathbf{z}|\boldsymbol{\theta}_i, \mathbf{x}) f(\mathbf{x}|\boldsymbol{\theta}_i) f(\boldsymbol{\theta}_i) d\boldsymbol{\theta}_i}{\int f(\mathbf{x}|\boldsymbol{\theta}_i) f(\boldsymbol{\theta}_i) d\boldsymbol{\theta}_i} \\ &= \frac{\int f(\mathbf{y}|\boldsymbol{\theta}_i) f(\boldsymbol{\theta}_i) d\boldsymbol{\theta}_i}{\int f(\mathbf{x}|\boldsymbol{\theta}_i) f(\boldsymbol{\theta}_i) d\boldsymbol{\theta}_i}. \end{aligned} \tag{2.7}$$

Since we have the same prior on the top and bottom of this last expression, the sensitivity to priors seen in the full marginal likelihood is much reduced.

Two variants of this basic idea are *intrinsic* and *fractional* Bayes factors. Intrinsic Bayes factors (Berger and Pericchi, 1996) use a subset $\mathbf{x}$ just large enough to ensure that $f(\boldsymbol{\theta}_i|\mathbf{x})$ is proper, and then average the resulting partial Bayes factors over all such subsets to remove the arbitrariness attendant on any particular choice of $\mathbf{x}$. The required averaging can be somewhat computationally costly. Hence *fractional* Bayes factors (O'Hagan, 1995) use the fact that if $b = \dim(\mathbf{x})/\dim(\mathbf{y})$, then $f(\mathbf{x}|\boldsymbol{\theta}_i) \approx f(\mathbf{y}|\boldsymbol{\theta}_i)^b$ (at least for large dimensions and exchangeable observations) and this approximation can be plugged into (2.7). Note that if the fractional Bayes factors are to select the right model in the large sample limit, then $b \to 0$ as the sample size tends to infinity. This consistency is automatic for intrinsic Bayes factors. Hence the Bayesian cross-validation approach of setting $\mathbf{x}$ to $\mathbf{y}$ with one datum omitted, and then averaging the results over each possible $\mathbf{z}$, will not give consistent model selection, but then neither does AIC (see Section 4.6). See Section 6.3.1 for fractional Bayes factor computation.

BIC: the Bayesian information criterion

An older approach, avoiding the difficulties of sensitivity to priors, is due to Schwarz (1978). Dropping the notation relating to particular models in the interests of clarity, the computation of the Bayes factor requires the evaluation of the marginal likelihood,

$$P = \int f(\mathbf{y}|\boldsymbol{\theta}) f(\boldsymbol{\theta}) d\boldsymbol{\theta},$$

for each model. Let n be the dimension of $\mathbf{y}$ and p be the dimension of $\boldsymbol{\theta}$, and define $f_p(\boldsymbol{\theta}) = f(\mathbf{y}|\boldsymbol{\theta})f(\boldsymbol{\theta})$ ($\mathbf{y}$ is the observed data vector here). Let $\tilde{\boldsymbol{\theta}}$ be the value of $\boldsymbol{\theta}$ maximising f_p. A Taylor expansion gives

$$\log f_p(\boldsymbol{\theta}) \simeq \log f_p(\tilde{\boldsymbol{\theta}}) - \frac{1}{2}(\boldsymbol{\theta} - \tilde{\boldsymbol{\theta}})^{\mathrm{T}} \left(-\frac{\partial^2 \log f_p}{\partial \boldsymbol{\theta} \partial \boldsymbol{\theta}^{\mathrm{T}}} \right) (\boldsymbol{\theta} - \tilde{\boldsymbol{\theta}})$$

$$\Rightarrow f_p(\boldsymbol{\theta}) \simeq f_p(\tilde{\boldsymbol{\theta}}) \exp\left\{ -\frac{1}{2}(\boldsymbol{\theta} - \tilde{\boldsymbol{\theta}})^{\mathrm{T}} \left(-\frac{\partial^2 \log f_p}{\partial \boldsymbol{\theta} \partial \boldsymbol{\theta}^{\mathrm{T}}} \right) (\boldsymbol{\theta} - \tilde{\boldsymbol{\theta}}) \right\}.$$

Recognising the term in $\{\}$ from the multivariate normal p.d.f., we have

$$P = \int f_p(\boldsymbol{\theta}) d\boldsymbol{\theta} \simeq f_p(\tilde{\boldsymbol{\theta}}) (2\pi)^{p/2} \left| -\frac{\partial^2 \log f_p}{\partial \boldsymbol{\theta} \partial \boldsymbol{\theta}^{\mathrm{T}}} \right|^{-1/2}.$$

Now assume, at least in the $n \to \infty$ limit, that $-\partial^2 \log f_p / \partial\boldsymbol{\theta}\partial\boldsymbol{\theta}^{\mathrm{T}} = n\boldsymbol{\mathcal{I}}_0$, where $\boldsymbol{\mathcal{I}}_0$ is a matrix such that $|\boldsymbol{\mathcal{I}}_0|$, is bounded above and below by finite positive constants, independent of n (and ideally close to 1). In the case of i.i.d. data, then $\boldsymbol{\mathcal{I}}_0$ is the (fixed) information matrix for a single observation. Under this assumption we have

$$\left| -\frac{\partial^2 \log f_p}{\partial\boldsymbol{\theta}\partial\boldsymbol{\theta}^{\mathrm{T}}} \right| = n^p |\boldsymbol{\mathcal{I}}_0|$$

and so

$$\log P \simeq \log f(\mathbf{y}|\tilde{\boldsymbol{\theta}}) + \log f(\tilde{\boldsymbol{\theta}}) + \frac{p}{2}\log(2\pi) - \frac{p}{2}\log n - \frac{1}{2}\log|\boldsymbol{\mathcal{I}}_0|.$$

Now as $n \to \infty$, $\tilde{\boldsymbol{\theta}} \to \hat{\boldsymbol{\theta}}$ (the MLE) while the terms that do not depend on n become negligible compared to those that do. So we arrive at

$$\mathrm{BIC} = -2\log f(\mathbf{y}|\hat{\boldsymbol{\theta}}) + p\log n \quad (\approx -2\log P).$$

Hence the difference in BIC between two models is a crude approximation to twice the log Bayes factor, and all other things being equal, we would select the model with the lowest BIC. Notice some arbitrariness here: there is really nothing in the preceding derivation to stop us from multiplying n in BIC by the finite positive constant of our choice. One interesting feature of BIC is that because it drops the prior, it is not susceptible to the problem with sensitivity to priors that affects the Bayes factor itself, but on the other hand the justification for dropping the prior seems somewhat artificial.

DIC: the deviance information criterion

In complex Bayesian models it is not always clear how to count the number of free parameters in the model. For example, the distinction between random effects and parameters in models is really terminological rather than formal in the Bayesian setting. This makes application of BIC problematic, as does BIC's dependence on knowledge of the posterior modes, which are not directly available via simulation.

In essence, the problem with counting free parameters is the introduction of priors. A prior restricts the freedom of a parameter to vary. In the limit in which the prior was a Dirac delta function, then the corresponding parameter would behave as a fixed constant, insensitive to the data, and should clearly not count in the tally of free parameters at all. Moving smoothly from this extreme to the other extreme of a fully uninformative prior, it seems reasonable that the contribution of the corresponding parameter to the free parameter count should increase smoothly from 0 to 1. This idea leads us to the notion of *effective degrees of freedom*.

Spiegelhalter et al. (2002) suggest a measure of the effective degrees of freedom of a Bayesian model that is readily computed from simulation output. They first define the *deviance* as

$$D(\boldsymbol{\theta}) = -2 \log f(\mathbf{y}|\boldsymbol{\theta}) + c,$$

where c is a neglectable constant depending only on the $\mathbf{y}$ (and hence not varying between models of a given $\mathbf{y}$). Using the notation $\bar{x}$ for 'mean of x', the proposed definition of the effective degrees of freedom is

$$p_D = \overline{D(\boldsymbol{\theta})} - D(\bar{\boldsymbol{\theta}}).$$

The definition is appealing because in the large sample limit in which the likelihood dominates the prior and the posteriors are approximately Gaussian, then $D(\boldsymbol{\theta}) - D\{E(\boldsymbol{\theta})\} \sim \chi^2_r$, by the same argument that leads to (2.4). But p_D is a direct estimate of $E[D(\boldsymbol{\theta}) - D\{E(\boldsymbol{\theta})\}]$, and $E(\chi^2_r) = r$. The deviance information criterion is then

$$\text{DIC} = D(\bar{\boldsymbol{\theta}}) + 2p_D,$$

which clearly has a somewhat similar form to AIC. Derivation of DIC in the context of approximately Gaussian posteriors is relatively straightforward, but it is applied much more widely, with some associated controversy. In any case the DIC cannot be justified if p_D is not small relative to the number of data in $\mathbf{y}$. It has the pragmatic appeal of being readily computable from simulation output, while being much less sensitive to the choice of vague priors than the marginal likelihood.

2.5.3 Interval estimation

The Bayesian approach to answering the question of what range of parameter values is consistent with the data is to pick out the range of values with high posterior probability. For example, a $\gamma 100\%$ *credible* set for θ_i is

$$\Omega = \{\theta_i : \int_\Omega f(\theta_i|\mathbf{y}) d\theta_i = \gamma; f(\theta_i \in \Omega|\mathbf{y}) > f(\theta_i \notin \Omega|\mathbf{y})\}$$

(i.e. it is the set containing the $\gamma 100\%$ of θ_i values with highest posterior probability). If the set is continuous then its endpoints define a $\gamma 100\%$ *credible interval*. Notice the difference from frequentist confidence intervals. Here the interval is fixed and the parameter is random, which is the opposite of the frequentist interpretation. Despite this difference, in the large sample limit, with informative data, Bayesian credible intervals and frequentist confidence intervals coincide.

2.5.4 Model checking

A particularly extreme Bayesian argument states that you should not check models, because doing so implies that you did not properly specify your uncertainty when setting up the Bayesian analysis and are therefore being 'incoherent'. This argument is somewhat impractical, and it is usually more pragmatic to view both models and priors as approximate representations of reality that it would be wise to check for gross infelicities. In part this checking can be done as in Section 2.4.4, but it is also wise to check the sensitivity of results to the specification of the prior, especially if it was chosen more or less arbitrarily, as is often the case. Simulation of replicate data implied by draws from the posterior distribution of the parameters can also be helpful, in order to check whether the posterior simulated data deviate in some systematic way from the observed data, indicating a problematic model (see e.g. Section 6.6.4).

2.5.5 The connection to MLE

We have already seen the large sample coincidence between posterior modes and MLEs and the large sample correspondence of Bayesian credible intervals and frequentist credible intervals. In fact, in many circumstances, in the large sample limit,

$$\boldsymbol{\theta}|\mathbf{y} \sim N(\hat{\boldsymbol{\theta}}, \boldsymbol{\mathcal{I}}^{-1}),$$

where $\hat{\boldsymbol{\theta}}$ and $\boldsymbol{\mathcal{I}}$ are as in (2.3).

2.6 Design

Statistical design theory is concerned with the design of surveys and experiments so as to obtain data that will be best able to answer the statistical questions of interest. This is a large topic with a rich theory, and much is known about how to go about producing practically useful designs. This section simply introduces two important design ideas.

The first key idea is *randomisation*. To make things concrete, consider the example of conducting an experiment to test whether a drug is effective at reducing blood pressure, relative to a standard treatment. In addition to the drug there are many other variables that may control blood pressure, such as age, percentage body fat, sex and so on. Because these factors are not of direct interest here, they are referred to as *confounding variables*. Strictly, a confounding variable is any variable that is associated with both the response variable *and* other predictor variables of interest, but in practice it is often difficult to rule out confounding for any variable that might affect the response. To know what the effect of the drug is, we must allow for the presence of the confounders. To see why, imagine we treated almost all women in the study with the new drug, and almost all the men with the standard treatment: now try to work out how you would disentangle the effect of sex from the effect of the drug. One solution is to measure the confounders and include them in the statistical model used to analyse the data. This is useful, but we cannot measure all possible confounders, because we do not even know what some of them might be. In the face of unmeasured confounders, how can we hope to make valid inferences about the effect of the drug?

The answer is *randomisation*. If patients are randomly allocated to drug type, then we break all possible association between the treatment the patient receives and the value of the confounding variables, so they cease to be confounders. In effect the part of the patient's blood pressure change that is due to these other variables can now be treated as patient-specific random error. This random error can easily be accounted for in the statistical modelling, and valid conclusions can be drawn about the drug's effect.

The key point is that the randomisation of experimental units (e.g. patients) to experimental treatments (e.g. drug treatment) turns the effects of unmeasured confounder variables into effects that can be modelled as random noise. It is this effect of randomisation that allows experiments to be used to test for *causal* effects in a way that is impossible with observational or survey data, where we cannot eliminate the systematic effects of unmeasured (and possibly unknown) confounders.

The second big idea in design is that we can adjust how experiments or surveys are done to improve the parameter estimates made using the resulting data. The idea is often to try to optimise some measure of the average variance of the parameter estimators (or of $\boldsymbol{\theta}|\mathbf{y}$ for a Bayesian analysis). If we are using maximum likelihood estimation then the approximate covariance matrix for $\hat{\boldsymbol{\theta}}$ is $\boldsymbol{\mathcal{I}}^{-1}$. The leading diagonal elements of this matrix give the approximate estimator variances. Two of the most common design criteria are then as follows:

1. A-optimality, which seeks to minimise the average estimator variance, which is equivalent to minimising $\text{tr}(\boldsymbol{\mathcal{I}}^{-1})$.

2. D-optimality, which seeks to minimise the determinant of the approximate covariance matrix, $|(\mathcal{I}^{-1}| = 1/|\mathcal{I}|$.

The idea is that the design is adjusted to minimise the chosen criterion. Sometimes this can be achieved analytically, but otherwise a numerical optimisation may be required. See Cox (1992) for an introduction to design.

2.7 Useful single-parameter normal results

The approximate normality of many estimators, as a result of the central limit theorem and large sample maximum likelihood results, means that some basic computations involving single normally distributed estimators are required repeatedly.

Suppose we know that $\hat{\theta} \sim N(\theta, \sigma_\theta^2)$, where σ_θ is known, but θ is not. We might want to test $H_0 : \theta = \theta_0$ versus $H_1 : \theta \neq \theta_0$, for some specified value θ_0. A moment's thought, or contemplation of the likelihood ratio statistic, leads to the test statistic

$$\frac{\hat{\theta} - \theta_0}{\sigma_\theta},$$

which will obviously have a $N(0, 1)$ distribution if H_0 is true.[9] Since the null distribution is symmetric, and large magnitude values of the statistic support H_1, the p-value is

$$p = \Pr\left(|Z| \geq \left|\frac{\hat{\theta} - \theta_0}{\sigma_\theta}\right|\right) \quad \text{where } Z \sim N(0, 1). \tag{2.8}$$

In obvious notation, here is some R code to compute this (three variants, giving identical results):

```
z.obs <- (theta.hat - theta.0)/sigma
pnorm(abs(z.obs),lower.tail=FALSE) + pnorm(-abs(z.obs))
pnorm(-abs(z.obs))*2        ## use symmetry of N(0,1)
pchisq(z.obs^2,lower.tail=FALSE,df=1) ## equivalent
```

In fact we seldom know σ_θ, and it is much more common that $\hat{\theta} \sim N(\theta, c^2\sigma^2)$, where c is a known constant and σ^2 is unknown but can be estimated by $\hat{\sigma}^2$. One option is just to plug $c\hat{\sigma}$ in place of σ_θ and use (2.8). This is fine in the large sample limit, but at finite sample sizes we can usually do a bit better by not ignoring the variability in $\hat{\sigma}^2$.

[9] With even shorter contemplation of the likelihood ratio we could equally have used $(\hat{\theta} - \theta_0)^2/\sigma_\theta^2$, which has a χ_1^2 distribution under H_0: the p-value is unchanged.

Suppose that $\hat{\sigma}^2$ is statistically independent of $\hat{\theta}$ and that $\hat{\sigma}^2/\sigma^2 \sim \chi^2_k/k$ for some positive integer k. In that case, from the definition of the t_k distribution (see Section A.1.3),

$$\frac{\hat{\theta} - \theta_0}{c\hat{\sigma}} \sim t_k.$$

The p-value computation for $\text{H}_0 : \theta = \theta_0$ versus $\text{H}_1 : \theta \neq \theta_0$ now uses

$$p = \text{Pr}\left(|T| \geq \left| \frac{\hat{\theta} - \theta_0}{c\hat{\sigma}} \right| \right) \quad \text{where } T \sim t_k.$$

In R something like the following would be used:

```
t.obs <- (theta.hat - theta.0)/(const*sigma.hat)
pt(-abs(z.obs),df=k)*2 ## use symmetry of t_k
```

The assumptions about $\hat{\sigma}^2$ may look restrictive, but there are quite wide classes of model estimation problems for which they hold. For example, they hold exactly in the case of linear regression models (see Chapter 7), and approximately for generalised linear models (in these cases k is the number of data less the number of estimated parameters, excluding $\hat{\sigma}^2$). Even when the conditions only hold approximately, use of t_k is usually an improvement on simply ignoring the variability in σ^2 and using $N(0, 1)$. In any case as $k \to \infty$, t_k tends to $N(0, 1)$.

Now consider confidence interval (CI) estimation, first in the known variance case. Suppose that we would accept H_0, above, for any θ_0 resulting in a p-value $\geq \alpha$. In that case we would have accepted all θ_0 values such that

$$z_{\alpha/2} \leq \frac{\hat{\theta} - \theta_0}{\sigma_\theta} \leq z_{1-\alpha/2},$$

where z_ϕ is the ϕ quantile of the $N(0, 1)$ distribution: the value such that $\text{Pr}(Z \leq z_\phi) = \phi$. By symmetry $z_{\alpha/2} = -z_{1-\alpha/2}$ ($z_{\alpha/2}$ will be negative). Rearrangement of the inequality leads to

$$\hat{\theta} + z_{\alpha/2}\sigma_\theta < \theta_0 < \hat{\theta} - z_{\alpha/2}\sigma_\theta$$

(i.e. a $(1 - \alpha)100\%$ confidence interval for θ is $\hat{\theta} \pm z_{\alpha/2}\sigma_\theta$). Here is one possible piece of R code, implementing this for a 95% CI:

```
theta.hat + qnorm(c(.025,.975))*sigma
```

When the variance is estimated, the derivation is identical, except that the t_k distribution is used in place of $N(0, 1)$. Letting $t_{k,\alpha/2}$ denote the

$\alpha/2$ quantile of t_k, the endpoints of the $(1-\alpha)100\%$ CI for θ become $\hat{\theta} \pm t_{k,\alpha/2} c\hat{\sigma}$. R code to implement this might be

```
theta.hat + qt(c(.025,.975),df=k)*const*sigma.hat
```

Although the results here all relate to single parameters, nothing in the preceding arguments requires that the model only contains one unknown parameter. θ could be a single element from a parameter vector.

Exercises

2.1 Find the maximum likelihood estimates of μ and σ for Example 1, Section 2.1. Find an exact 95% confidence interval for μ. Compare it to the approximation based on (2.3). Compute an approximate confidence interval for σ.

2.2 By appropriate use of the `qnorm`, `sort`, `plot` and `abline` functions in R, check the model fit produced in question 2.1. Is the model adequate?

2.3 Using R, produce a contour plot of the log likelihood for the second model for the temperature data in Section 2.1, against μ and σ for $\alpha = 3$. Approximately find the MLE of μ and σ for the given α.

2.4 Using R, write a function to evaluate the log likelihood of θ_l for Example 4 in Section 2.1 (hint: see `?dexp`). Plot the log likelihood against θ_l over a suitable range, and by making use of (2.4) and the definition of a confidence interval, find a 95% confidence interval for θ_l (`pchisq` is also useful).

2.5 Write an R function to evaluate the log likelihood of model (2.2) in Section 2.2 by making use of the `chol` function in R (see Section B.2).

2.6 Consider simulated data `x <-rnorm(10)+1`, for which the Bayesian model $x_i \sim N(\mu, 1)$, $\mu \sim U(-k, k)$ (i.e. the prior for μ is a uniform distribution on $[-k, k]$). Making use of the R function `integrate`, investigate the sensitivity of the marginal likelihood for this model, when k is changed from 2 to 20. `Vectorize` may be useful for converting your joint density function to a form suitable for `integrate`.

2.7 Show that if independent observations x_i have an exponential distribution with parameter λ, and λ has a gamma distribution as a prior, then the posterior distribution of λ is also a gamma distribution (see Section A.2.2).

3

R

Statistical analysis of interesting datasets is conducted using computers. Various specialised computer programmes are available to facilitate statistical work. For using general statistical theory directly with custom-built models, R is probably the most usefully flexible of such programmes.

R (R Core Team, 2012) is a progamming language and environment designed for statistical analysis. It is free (see `http://cran.r-project.org` to obtain a copy) and is written and maintained by a community of statisticians. A major design feature is extendibility. R makes it very straightforward to code up statistical methods in a way that is easy to distribute and for others to use. The first place to look for information on getting started with R is `http://cran.r-project.org/manuals.html`. I will assume that you have installed R, can start it to obtain a command console, and have at least discovered the function `q()` for quitting R.[1]

The following web resources provide excellent guides to the R language at different levels.

- `http://cran.r-project.org/doc/contrib/Short-refcard.pdf` is a four page summary of key functions and functionality.
- `http://cran.r-project.org/doc/contrib/R_language.pdf` is a very concise introduction to and reference for the structure of the language.
- `http://cran.r-project.org/doc/manuals/R-lang.html` is the main reference manual for the language.

A huge amount of statistical functionality is built into R and its extension packages, but the aim of this chapter is simply to give a brief overview of R as a statistical programming language.

[1] When you quit R, it will ask you if you want to save the workspace image. If you reply 'yes' then all the objects created and not subsequently destroyed in your session will be saved to disk and reloaded next time you start R. Usually you do not want to do this.

3.1 Basic structure of R

When you start R (interactively) two important things are created: a command prompt at which to type commands telling R what to do, and an *environment*, known interchangeably as the 'global environment' or 'user workspace' to hold the objects created by your commands. Unlike the command prompt, you do not see the global environment directly, but it is there as an extendible chunk of computer memory for holding your data, commands and other objects.

Generically in R an 'environment' consists of two things. The first, known in R jargon as a *frame*, is a set of symbols used to refer to objects, along with the data defining those objects. The second is a pointer to an enclosing environment. As we will see, R makes use of different environments arranged in a tree structure when organising the evaluation and manipulation of objects. In a slightly Zen manner, the base environment of the tree contains nothing at all. For the most part environments act as seamless computational infrastructure that the programmer is largely unaware of, but for some purposes it is important to know about them.

Everything in R is an object living in an environment, including R commands themselves. Here is a line of R code to create an object called 'a' and to assign it the value 2 (using the assignment operator `<-`):

```
> a <- 2
```

As soon as I press return, the text "`a <-2`" is sent to the *parser* to be checked for correctness (i.e. whether it is a valid statement in R) and to be converted to an internal representation for evaluation, known as an *expression* object. The expression is then evaluated, which has the effect of creating an object in the user workspace referred to by the symbol `a` and containing the single number 2.

Once an object is created, it can be referred to by name and used to create other objects. For example,

```
> b <- 1/a
```

Having created objects you often need to check their contents. Just typing the name of an object causes R to print it (actually to call the `print` function for the class of object concerned):

```
> b
[1] 0.5
```

`ls()` lists all the objects in the global environment and `rm(b)` would remove the object called `b`.

R is a functional programming language: it is structured around functions that take objects as arguments and produce other objects as results. Even basic operators such as + and * are actually implemented as functions. Within R we can create objects that are functions. Suppose we want to create a function foo that takes arguments a and b and returns the value of $b \log(a) - 1/2$. Here is how to define such a function object:

```
foo <- function(a,b) {
  b * log(a) - 0.5
}
```

The curly brackets { and } enclose the R commands defining how the arguments a and b are converted into the result returned by the function. Whatever is evaluated on the last line of the function is taken to be its return value. So

```
> foo(2,3)
[1] 1.579442
```

prints the value of foo evaluated at $a = 2, b = 3$.

R evaluates commands once they appear complete and a line end has been encountered. Commands can be split over several lines, but you then need to be careful that they could not be interpreted as being complete at the end of one of the lines, before they really are. Conversely, if several complete commands are to be included on a single line then they must be separated by ';'. Commands can be grouped using curly brackets, { and }. Once you have started a group of commands with a {, it will not be complete and ready to be parsed and evaluated until you close it with a }.

You will have noticed from this discussion that R is an interpreted language. Commands are interpreted and executed as they are encountered (rather than being converted en masse into binary instructions and then executed, as in a compiled language, such as C). This has two important consequences. First, we will have to worry about achieving efficiency in repetitive tasks, to ensure that interpretation of what we want R to do does not take longer than actually doing it. Second, it means that it is possible to write R code that itself writes R code and runs it.

3.2 R objects

Objects in R are either language objects of some sort or are the result of a function call. Therefore, in contrast to many programming languages, we do not need to explicitly declare the type of a variable before using it: the type is determined by the function creating the variable. There are a number of basic object types (*classes*) in R, of which the most important for data handling are vectors and arrays. *Lists* are used to create objects containing several different sorts of object.

As well as their class, objects in R carry around information about the basic type of thing that they are made up of. Somewhat confusingly, they actually carry around three different classifications of the sort of basic thing they are made up of: their *type*, *mode* and *storage mode*. The following code illustrates this by creating the vector $(1, 2, 3, 4)$ and examining its type, mode and storage mode:

```
> b <- 1:4
> typeof(b)
[1] "integer"
> mode(b)
[1] "numeric"
> storage.mode(b)
[1] "integer"
```

Usually it is not necessary to worry much about the modes and type of an object: for example, the conversion between real and integer numbers is automatic and need seldom concern the programmer. The exception is when calling code written in other languages, when it is essential to know the storage mode of data being passed to the external code from R.

Objects can also carry a variety of extra information as *attributes*. Attributes have a name and can be an object of any type. They behave rather like a whole bunch of notes stuck onto the object and carried around with it. The `attributes` function lets you access all the attributes of an object, whereas the `attr` function allows individual attributes to be set and extracted. As an example, let's give the vector `b`, above, an attribute consisting of a 2×2 matrix (and then print it):

```
> attr(b,"mat") <- matrix(1:4,2,2)
> attr(b,"mat")
     [,1] [,2]
[1,]    1    3
[2,]    2    4
```

You can add as many attributes as you like, and they are used by R itself in many ways, including in the implementation of matrices and higher dimensional arrays. The class of an object is somewhat like a special attribute and is used in R's basic *object orientation* mechanism (see Section 3.6).

Here are the basic sorts of objects that are needed for manipulating data in R. For a complete list see the sources listed at the start of the chapter.

- *Vectors* are the default structures for storing real, complex, integer, logical and character data. Scalars are simply vectors of length 1. Here is some code to create a numeric vector `d`, check how many elements it has, and print the third element:

```
> d <- c(1,3.56,9)
> length(d)
[1] 3
> d[3]
[1] 9
```

- *Arrays* are vectors with a `dim` attribute and are of class `"array"`. The following creates a three dimensional array, displays its `dim` attribute, and prints its element $2, 1, 3$:

```
> b <- array(1:24,c(2,3,4))
> attributes(b)
$dim
[1] 2 3 4
> b[2,1,3]
[1] 14
```

  Array elements can be accessed by providing an index for each dimension, as just shown, or by providing a single index for accessing elements of the underlying vector. Arrays are stored in the underlying vector in 'column major' order, so if d is the `dim` attribute $b[i, j, k]$ is equivalent to $b[i + (j - 1)d_1 + (k - 1)d_1 d_2]$; that is `b[2,1,3]` refers to the same location as `b[14]`, in this case.
- *Matrices* are two dimensional arrays of class `"matrix"`. They are treated as a separate class to facilitate numerical linear algebra with matrices.
- *Factors* are, conceptually, vectors of labels that serve to group other data. They have a special place in statistical modelling (see e.g. Chapter 7) and as such require special handling. In R, factors have class `"factor"` and another attribute `"levels"`, which is a vector of the unique labels occurring in the factor object. If you print a factor variable, then what is printed is the label given in each element of the vector. However, what is actually stored is a set of integers indexing the `"levels"` attribute, and it is the print function that is actually doing the conversion from stored integer to corresponding label.
- *Data.frames* are matrices of data, where each column has a name, but not necessarily the same type. (e.g. having a mixture of logical, numeric, factor and character columns is no problem). This format is a natural way of storing statistical datasets. Here is a short example:

```
> dat <- data.frame(y=5:7,lab=c("um","er","er"))
> dat
  y lab
1 5 um
2 6 er
3 7 er
```

By default the character vector `lab` was converted to a factor variable.[2] Data frames can be accessed like any other matrix, but also by variable name and as a list (see next). In consequence, `dat[2,1]`, `dat$y[2]` and `dat[[1]][2]` all access the same entry (6).

- *Lists* are the basic building blocks for all sorts of complicated objects in R. Lists consist of any number of numbered, and optionally named, items, each of which can be an object of any type. For example,

  ```
  > li <- list(a="fred",1:3,M=matrix(1,2,2))
  ```

  Elements can be accessed by number, starting from 1, using double square brackets (e.g. `li[[1]]` accesses `"fred"`). If the item has a name then this provides an alternative access method using `$`. For example, `li$a` also accesses `"fred"`.

3.3 Computing with vectors, matrices and arrays

Data manipulation in R is vector based. That is, wherever possible we work with whole vectors, rather than with individual elements of vectors, because the former is much more efficient in an interpreted language. So standard operators and mathematical functions in R are defined in a vector oriented manner, as best illustrated by an example.

Suppose we have vectors $\mathbf{x}$ and $\mathbf{y}$ and want to evaluate the vector $\mathbf{z}$, the elements of which are defined as $z_i = \sin(x_i)y_i - y_i^{x_i}/\exp(x_i)$. If `x` and `y` are the vectors in R, then

```
> z <- sin(x)*y - y^x/exp(x)
```

computes $\mathbf{z}$. The key point is that the functions and operators are operating *elementwise* on the vector components.

There are built-in functions for a number of common vector operations that are not purely elementwise. For example:

`sum(x)` to evaluate $\sum_i x_i$.
`prod(x)` to evaluate $\prod_i x_i$.
`cumsum(x)` to evaluate $z_i = \sum_{j=1}^{i} x_j$.
`cumprod(x)` to evaluate $z_i = \prod_{j=1}^{i} x_j$.

3.3.1 The recycling rule

When working with vectors we often want to perform operations involving a scalar and a vector (such as multiplying all the elements of a vector by

[2] See `?data.frame` for how to turn off this conversion.

two). In R, scalars are simply vectors of length one: there is nothing special about them. However, R has a *recycling rule*, which is a vector-oriented generalisation of what happens when a scalar is multiplied by a vector. The recycling rule states that when two vectors of different length appear in an operation, then the shorter is replicated as many times as is necessary to produce a vector of the length of the longer vector, and this recycled vector is what is used in computation.

So conceptually if `x <-c(1,2,3)`, then `z <-2*x` results in `2` being recycled three times to produce a vector of three `2`s, which is then multiplied by `x` elementwise to produce `z`. Here is an example of recycling in action:

```
> a <- 1:4
> b <- 1:2
> a + b
[1] 2 4 4 6
```

Recycling can be very useful once you get used to it. For example, suppose that we want to form $\mathbf{A} = \mathbf{WX}$ where $\mathbf{W}$ is a diagonal matrix with diagonal elements $\mathbf{w}$, and $\mathbf{X}$ is some $n \times n$ matrix. One option for doing this is to form $\mathbf{W}$ explicitly and then multiply $\mathbf{X}$ by it:

```
W <- diag(w); A <- W%*%X
```

This uses something like $2n^3$ arithmetic operations, most of which involve products with zero, and make no contribution to the final result. But recalling that matrices are actually stored columnwise in vectors, we can simply exploit the recycling rule to compute the result with n^2 operations and no wasted products with zero:

```
A <- w * X
```

R will produce a warning if the number of elements in the longer vector is not an integer multiple of the number of elements in the shorter vector. It will also refuse to recycle vectors with a dimension attribute.

3.3.2 Matrix algebra

Clearly, vectorised elementwise operations mean that `A*B` does not perform matrix multiplication and `A/B` does not produce $\mathbf{AB}^{-1}$. Instead these and other matrix operations have special functions and operators:

- `%*%` is the matrix multiplication operator. `A %*%B` performs the matrix multiplication $\mathbf{AB}$, checking for compatible dimensions. This is also used for matrix-vector multiplication.
- `%x%` is the Kronecker product. So `A %x% B` produces $\mathbf{A} \otimes \mathbf{B}$.

- `t(A)` returns the transpose of its argument.
- `solve(A,B)` forms $\mathbf{A}^{-1}\mathbf{B}$. To form $\mathbf{A}\mathbf{B}^{-1}$ use `t(solve(t(B),t(A)))` since $\mathbf{A}\mathbf{B}^{-1} = (\mathbf{B}^{-\mathrm{T}}\mathbf{A}^{\mathrm{T}})^{\mathrm{T}}$. `solve(A)` returns $\mathbf{A}^{-1}$, but should rarely be used, because it is usually more expensive and less stable to compute the inverse explicitly.
- `crossprod(A)` produces $\mathbf{A}^{\mathrm{T}}\mathbf{A}$ at least twice as quickly as `t(A)%*%A` would do.

When computing with these basic operations you need to be very careful about the ordering of operations. An interesting feature of numerical linear algebra is that many expressions can be evaluated in a number of different orders, all of which give the same result, but can differ in their computational speed by several orders of magnitude. To see this, consider computing $\mathbf{BCy}$, where, from left to right the matrix dimensions are $n \times m$, $m \times n$ and $n \times 1$ (a vector). R function `system.time` lets us examine the effect of operation ordering, while n and m are set to 1000 and 2000.

```
> system.time(z <- B%*%C%*%y)
   user  system elapsed
  2.706   0.009   2.720
> system.time(z <- B%*%(C%*%y))
   user  system elapsed
  0.013   0.000   0.013
```

Both lines compute the same quantity here, but the second is *much* faster. Why? In the first case R simply evaluates the expression left to right: the matrix produced by $\mathbf{BC}$ is formed first at the cost of $2n^2m$ arithmetic operations, after which the result is multiplied by $\mathbf{y}$ at a further cost of $2n^2$ operations. In the second case the brackets force R to compute the vector $\mathbf{Cy}$ first using $2mn$ operations, and then to multiply it by $\mathbf{B}$ at the cost of $2mn$ more operations. So the latter approach involves something like a factor of n fewer operations.[3] This is clearly an issue that can not be ignored, but it is also rather straightforward to deal with (see Appendix B).

Functions `chol`, `qr`,[4] `eigen` and `svd` produce the Choleski, QR, eigen and singular value decompositions of their arguments, respectively. In the case of `qr` the decomposition is returned in a compact form, which can then be manipulated with helper functions (see `?qr`). `forwardsolve` and `backsolve` are versions of `solve` to use with, respectively, lower and upper triangular first arguments: for an $n \times n$ argument they are a factor of n more efficient than using `solve` in these cases. Functions `det` and

[3] We do not see quite that speed up here because of other overheads in both calculations.
[4] Beware of the default `tol` argument for `qr` in R: it is set quite high for some purposes.

determinant are also provided for evaluating determinants, but for many statistical applications determinants should rather be computed directly from the triangular QR or Choleski factor, which is likely to be needed anyway. Appendix B provides more information on matrix decompositions.

Functions ncol and nrow return the number of rows or columns of their argument, whereas rowSums and colSums return vectors of sums for each row or each column, respectively. kappa efficiently estimates the condition number of a matrix (see Section B.3.3), and norm computes various matrix norms. apply and its relatives, covered next, are also useful with matrices.

Matrices consisting mainly of zeroes are known as *sparse matrices*. The Matrix package supplied with R provides the facilities for working with sparse matrices, but be warned that you need to understand pivoting and the issue of *infil* in order to make good use of it (see Davis, 2006).

3.3.3 Array operations and apply

Beyond matrices, many array operations are accomplished using vector arithmetic and the indexing and subsetting facilities to be covered next. However, there are two common array-oriented tasks that deserve special mention: applying a function to some margins of an array, using the apply function, and forming array products by 'summing over array indices', using Jonathon Rougier's tensor[5] package.

The apply function takes a single array as its first argument, a vector of dimension indices as the next argument, and then a function to be applied to the data given by the indices. Here is a simple example using apply to sum over the rows and columns of a 2×3 matrix (a two dimensional array).

```
> A <- matrix(1:6,2,3);A
     [,1] [,2] [,3]
[1,]    1    3    5
[2,]    2    4    6
> apply(A,1,sum)
[1]  9 12
> apply(A,2,sum)
[1]  3  7 11
```

The first call to apply specifies that the sum function should be applied to each row (the first dimension) of A in turn, and the results returned. The second call applies sum to each column (the second dimension) of A. To be

[5] In physics and geometry a *vector* has a magnitude and an associated direction, requiring a one-dimensional array of numbers to represent it. A *tensor* has a magnitude and d associated directions, and it requires a d-dimensional array of numbers to represent it.

clear about what `apply` is doing, it helps to see what happens if we specify rows and columns in the second argument:

```
> apply(A,c(1,2),sum)
     [,1] [,2] [,3]
[1,]    1    3    5
[2,]    2    4    6
```

So `apply` has taken the data given by each combination of row and column index (a single number in the case of a two dimensional array) and applied `sum` to that, giving the matrix result shown, which is just the original `A` here.

`apply` can be used in the same way with arrays of any dimension: see `?apply` for more. Furthermore there are versions of `apply` for lists (see `?lapply`), and for applying functions over subsets of vectors (see `?tapply`). Related functions `aggregate` and `sweep` are also useful.

Now consider array products. We have a c-dimensional array A and a d-dimensional array B and want to find the array that results by forming inner products of some of their dimensions. For example,

$$C_{ipqvw} = \sum_{jkl} A_{ijklpq} B_{kjlvw}.$$

If we write some indices as superscripts, then this can be written more compactly using Einstein's summation convention as $C^{ipq}_{vw} = A^{ijklpq} B_{kjlvw}$. The idea is that we sum over the product of the elements given by shared indices. Here is a concrete example forming $A^{ijk} B_{kjl}$ in R:

```
> A <- array(1:24,c(2,3,4))
> B <- array(1:72,c(4,3,5))
> require(tensor) ## load the tensor library
> tensor(A,B,c(2,3),c(2,1))
     [,1] [,2] [,3] [,4] [,5]
[1,] 1090 2818 4546 6274 8002
[2,] 1168 3040 4912 6784 8656
```

`tensor` takes the arrays as arguments, followed by two vectors giving the dimensions to be summed over.

3.3.4 *Indexing and subsetting*

Operations often have to be applied to only a subset of a vector or array or to only some dimensions of an array. To facilitate this efficiently, R has a rich variety of indexing and subsetting operations that can be applied to arrays and vectors. First let us consider vectors.

Vectors can be indexed using a vector of integers giving the locations of the required elements within the indexed vector, or by a logical array of the same length as the indexed vector, having a `TRUE` for each required element. Here is an example:

```
> x <- c(0,-1,3.4,2.1,13)
> ii <- c(1,4,5)
> x[ii]
[1] 0.0 2.1 13.0
> il <- c(TRUE,FALSE,FALSE,TRUE,TRUE)
> x[il]
[1] 0.0 2.1 13.0
```

The index version is somewhat more flexible than the logical version, in that elements can be selected more than once and in any order. For example,

```
> ii <- c(5,4,1,1)
> x[ii]
[1] 13.0 2.1 0.0 0.0
```

However, the logical version is better for extracting values according to some condition. For example, the values of `x` less than 2.5 can be extracted as follows:

```
> il <- x < 2.5
> il
[1] TRUE TRUE FALSE TRUE FALSE
> x[il]
[1] 0.0 -1.0 2.1
```

or with the single command `x[x < 2.5]`. It is often helpful to convert the logical version to the index version, and the `which` function does this:

```
> ii <- which(x < 2.5); ii
[1] 1 2 4
```

(`ii <-(1:5)[x < 2.5]` is equivalent).

Index vectors of either type can also appear on the 'left-hand side' of an assignment[6]. This example resets any element of `x` to 1 if it is less than 2:

```
> x[x < 2] <- 1; x
[1] 1.0 1.0 3.4 2.1 13.0
```

The examples so far have involved only simple conditions, but often more complicated subsetting is required. This can be achieved using elementwise logical 'or' and 'and' operators, '`|`' and '`&`'. For example, consider selecting the elements of a vector `z` that are between -1 and 2:

```
> z <- c(3.6,-1.2,1,1.6,2,20)
> z[z >= -1 & z <= 2]
[1] 1.0 1.6 2.0
```

[6] Actually the assignment arrow can point in either direction, so this really means 'at the pointy end of the assignment operator'.

`z[z < -1 | z > 2]` extracts the complement of this subset, as does the alternative `z[!(z >= -1 &z <= 2)]`, which uses the 'not' operator, '`!`'.

Another common task is to apply some function to all of several non-overlapping subsets of a vector, a task which `tapply` accomplishes. The vector of data, `X`, is its first argument, followed by a vector or list of vectors, `INDEX`, containing one or more factor variables, each of the same length as `X`. All elements of `X` sharing the same combination of factor levels from `INDEX` are in the same group, and the subvectors containing these groups supply the argument to `tapply`'s third argument, the function `FUN`. For example, suppose that the means of the first two, the next three and the final element of `z` are required:

```
> fac <- factor(c(1,1,2,2,2,3))
> tapply(z,fac,mean)
       1        2         3
 1.200000 1.533333 20.000000
```

Matrices and arrays generally require one further sort of subsetting: the extraction of particular rows and columns. Actually this works by making use of the fact that the absence of an indexing array is taken to mean that the whole vector is required. For example, `x` and `x[]` both return the whole vector, `x`. Similarly, `X[i,]` and `X[,j]` extract, respectively, row `i` and column `j` of matrix `X`.

Indexing vectors and missing indices can be mixed in any way you like; for example,

```
> a <- array(1:24,c(2,3,4))
> a[1,,2:3]
     [,1] [,2]
[1,]    7   13
[2,]    9   15
[3,]   11   17
```

Notice, however, that the task of extracting scattered elements of a matrix or array is more difficult. Suppose I want to extract the three elements $(1, 3)$, $(4, 2)$ and $(2, 1)$ from a 4×3 matrix `B`. Naively I might try

```
> B <- matrix(1:12,4,3)
> i <- c(1,4,2); j <- c(3,2,1)
> B[i,j]
     [,1] [,2] [,3]
[1,]    9    5    1
[2,]   12    8    4
[3,]   10    6    2
```

... not what is required at all (but completely consistent with the preceding description of how indexing works). Here the underlying vector storage of arrays comes to the rescue. Recalling that arrays are stored in column major order, we can create an appropriate vector indexing the underlying vector storage in order to extract the required elements:

```
> B[i+(j-1)*4]
[1] 9 8 2
```

One important detail of array subsetting is that if a subset results in a single vector, then dimension attributes are dropped by default. This can cause problems in code designed for the case where we do not know in advance whether an array or a vector will be returned by an operation. In this case we can force dimension attributes to be retained, as follows:

```
> B[1,] ## vector result
[1] 1 5 9
> B[1,,drop=FALSE] ## 1 by 3 matrix result
     [,1] [,2] [,3]
[1,]    1    5    9
```

3.3.5 Sequences and grids

Many computations require the production of regular sequences of numbers (or occasionally other variables). The simplest is a sequence of numbers incrementing or decrementing by 1. `a:b` produces a sequence starting at `a` and finishing at `a+k` where `k` is the largest integer such that $\texttt{a}+\texttt{k} \le \texttt{b}$ if $\texttt{a} < \texttt{b}$ or such that $\texttt{a} - \texttt{k} \ge \texttt{b}$ otherwise. Usually `a` and `b` are integers. For example,

```
> i <- 1:10; i
 [1]  1  2  3  4  5  6  7  8  9 10
```

Function `seq` produces sequences with increments that need not be unity. Its first two arguments specify endpoints, while argument `by` specifies an increment to apply, or alternatively `length` specifies how many elements the sequence should have. For example

```
> x <- seq(0,1.5,length=4); x
[1] 0.0 0.5 1.0 1.5
```

Frequently, sequences should repeat in some way and `rep` facilitates this. Its first argument is a 'base sequence', and the second argument specifies how its elements are to be repeated. Here are some examples:

```
> rep(x,2)      ## whole sequence repeat
[1] 0.0 0.5 1.0 1.5 0.0 0.5 1.0 1.5
> rep(x,each=2) ## element-wise repeat
[1] 0.0 0.0 0.5 0.5 1.0 1.0 1.5 1.5
> rep(x,rep(2,4)) ## element-wise (more flexible)
[1] 0.0 0.0 0.5 0.5 1.0 1.0 1.5 1.5
```

For the last form, the second argument is a vector of the same length as `x` specifying how many times each element of `x` is to be repeated (its elements may differ, of course).

Regular sequences in more than one dimension are also useful: grids. One option for generating a grid is to use `rep`. However, it is often easier to use the function `expand.grid`, which takes named arguments defining the mesh points in each dimension, and returns a data frame with columns corresponding to each margin, expanded so that the points fall on a regular grid. For example,

```
> z <- seq(-1,0,length=3)
> expand.grid(z=z,x=x)
     z   x
1 -1.0 0.0
2 -0.5 0.0
3  0.0 0.0
4 -1.0 0.5
.   .   .
12 0.0 1.5
```

Any dimension of grid can, in principle, be generated. Often a grid is generated in order to evaluate a function of several variables over some domain. In that case it can be more convenient to use the function `outer` which generates the evaluation grid internally, and returns the function values, evaluated on the grid, as an array.

3.3.6 Sorting, ranking, ordering

`sort` will return its argument sorted into ascending order (set the second argument `decreasing` to `TRUE` to get descending order). For example,

```
> set.seed(0); x <- runif(5); x
[1] 0.8966972 0.2655087 0.3721239 0.5728534 0.9082078
> sort(x)
[1] 0.2655087 0.3721239 0.5728534 0.8966972 0.9082078
```

Often it is necessary to apply the reordering implied by sorting one variable to some other variables. `order` will return an appropriate index vector for doing this (see also `?sort.int`), illustrated here by resorting `x` itself:

```
> io <- order(x); io
[1] 2 3 4 1 5
> xs <- x[io]; xs
[1] 0.2655087 0.3721239 0.5728534 0.8966972 0.9082078
```

A related task is to find the rank of data in a vector, and `rank` does this. In the absence of ties, then `rank` is the inverse function of `order` in the following sense:

```
> ir <- rank(x); ir
[1] 4 1 2 3 5
> xs[rank(x)]
[1] 0.8966972 0.2655087 0.3721239 0.5728534 0.9082078
```

Another way of 'inverting' `io` is with

```
> um <- rep(0,5)
> um[io] <- 1:5; um
[1] 4 1 2 3 5
```

Similarly, using `ir` on the left-hand side results in `um` being `io`. This sort of construction is useful when dealing with matrix pivoting (as used, optionally, in `qr` and `chol`, for example).

3.4 Functions

Functions were introduced in Section 3.1, but some more detail is required to write them effectively. Formally a function consists of an argument list, a body (the code defining what it does), and an environment (which is the environment where it was created). Generally, functions take objects as arguments and manipulate them to produce an object, which is returned. There are two caveats to this general principle.

1. A function may have side effects, such as printing some output to the console or producing a plot. Indeed a function may only produce a side effect, and no return object. Generally side effects that modify objects that are external to the function are to be avoided, if code is to be clean and easy to debug.
2. A function may make use of objects not in its argument list: if R encounters a symbol not in the function argument list and not previously created within the function, then it searches for it, first in the environment in which the function was *defined*[7] (which is not necessarily the environment from which it was called). If that fails it looks in the environments returned by function `search()`. A benign use of this mechanism is to call other functions not in a function's argument list, or to access constants such as those stored in `.Machine`. Using this

[7] This is known as 'lexical scoping', because the parent environment of the function is where it was written down.

mechanism to provide a function with other objects that you have created is generally bad practice, because it makes for complex hard-to-debug code. Generally all objects that a function needs should be provided as its arguments. If this gets unwieldy, then group the arguments into a smaller number of list arguments.

Here is an example of a function definition. It generalises one-to-one real functions with power series representations to symmetric matrices:

```
mat.fun <- function(A,fun=I) {
  ea <- eigen(A,symmetric=TRUE)
  ea$vectors %*% (fun(ea$values)*t(ea$vectors$))
}
```

'`function(A,fun=I)`' indicates that a function is to be created with arguments `A` and `fun`. In this case the function created is given the name `mat.fun`, but functions are sometimes used without being given a name (for example, in the arguments to other functions). The argument list gives the names by which the function arguments will be referred to within the function body. Arguments may be given default values to be used in the event that the function is called without providing a value for that argument. This is done using `name = default` in the argument list. `fun=I` is an example of this, setting the default value of `fun` to the identity function.

Next comes the body of the function given by the R expressions within the curly brackets `{ ... }` (if the function body consists of a single expression, then the brackets are not needed). The function body can contain any valid R expressions. The object created on the last line of the function body is the object returned by the function. Alternatively the object can be returned explicitly using the `return` function. For `mat.fun`, the eigen decomposition of the first argument is obtained and then used to produce the generalised version of `fun`.

Now let us use the function, with a random matrix. First a sanity check:

```
> set.seed(1)
> m <- 3; B <- crossprod(matrix(runif(m*m),m,m))
> B; mat.fun(B)
          [,1]      [,2]      [,3]
[1,] 0.5371320 0.8308333 0.8571082
[2,] 0.8308333 1.6726210 1.5564220
[3,] 0.8571082 1.5564220 1.7248496
          [,1]      [,2]      [,3]
[1,] 0.5371320 0.8308333 0.8571082
[2,] 0.8308333 1.6726210 1.5564220
[3,] 0.8571082 1.5564220 1.7248496
```

which confirms that the default behaviour is to return the first argument.

Now consider what actually happened when the function was called (by `mat.fun(B)`). R first matches the arguments of the function to those actually supplied, adopting a rather permissive approach to so doing. First it matches on the basis of exact matches to argument names ('`A`' and '`fun`' in the example). This does not mean that R is looking for `B` to be called `A` in the example; rather it is looking for statements of the form `A=B`, specifying unambiguously that object `B` is to be taken as argument '`A`' of `mat.fun`. After exact matching, R next tries partial matching of names on the remaining arguments; for example `mat.fun(B, fu=sqrt)` would cause the `sqrt` function to be taken as the object to be used as argument `fun`. After matching by name, the remaining arguments are matched by position in the argument list: this is how R has actually matched `B` to `A` earlier. Any unmatched argument is matched to its default value.

R next creates an *evaluation frame*: an extendible piece of memory in which to store copies of the function arguments used in the function, as well as the other objects created in the function. This evaluation frame has the environment of the function as its parent (which is the environment where the function was defined, remember).

Having matched the arguments, R does not actually evaluate them immediately, but waits until they are needed to evaluate something in the function body: this is known as *lazy evaluation*. Evaluation of arguments takes place in the environment from which the function was called, except for arguments matched to their default values, which are evaluated in the function's own evaluation frame.

Preliminaries over, R then evaluates the commands in the function body, and returns a result.

Notice that arguments are effectively copied into the function's evaluation frame, so nothing that is done to a function argument within the function has any effect on the object that supplied that argument 'outside' the function. Within the body of `mat.mod` argument `A` could have been replaced by some poetry, and matrix `B` would have remained unaltered.

Here is an example of calling `mat.mod` to find a matrix inverse:

```
> mat.fun(A = B, fun = function(x) 1/x)
          [,1]      [,2]      [,3]
[1,] 10.108591 -2.164337 -3.070143
[2,] -2.164337 4.192241 -2.707381
[3,] -3.070143 -2.707381 4.548381
```

In this case both arguments were supplied by their full name, and a function definition was used to supply argument `fun`.

3.4.1 The '...' argument

Functions can also have a special argument '...', which is used to create functions that can have variable numbers of arguments. It is also used to pass arguments to a function that may in turn be passed on to other functions, without those arguments having to be declared as arguments of the calling function: this is useful for passing arguments that control settings of plotting functions, for example.

Any arguments supplied in the call to a function, that are not in the argument list in the function definition, are matched to its '...' argument, if it has one.[8] The elements of '...' can be extracted into a list. The following simple function's only purpose is to do this, and thereby show you all you need to know to work with '...':

```
dum <- function(...) {
  arg <- list(...)
  arg.names <- as.list(substitute(list(...)))[-1]
  names(arg) <- arg.names
  arg
}
```

The first line of the function body extracts the arguments and puts them in a list. The next line extracts the names of the arguments (in the calling environment, obviously). Look up `?substitute` to understand exactly how it works. The names are then given to the elements of the list. Here it is in unexciting action, with just two arguments:

```
> a <- 1; b <- c("um","er")
> dum(a,b)
$a
[1] 1
$b
[1] "um" "er"
```

As mentioned, a major use of '...' is to pass arguments to a function for it to pass on to another function. R's optimisation function `optim` uses this mechanism to pass arguments to the function that it is minimising. `optim` is designed to minimise functions with respect to their first argument (a vector). The function to be optimised may have many other arguments, of no concern to `optim`, except that values for them have to be supplied. `optim` does not 'know' what these are called nor how many of them there are: it does not need to because they can be provided as named arguments matched to '...' and passed to the function that way. For example, here is a function for `optim` to minimise, and the call to do so:

[8] This has the slightly unfortunate side effect that mistyped argument names do not generate obvious warnings.

```
ff <- function(x,a,b) {
  (x[1]-a/(x[2]+1))^2 + (x[2]-b/(x[1]+1))^2
}
optim(c(0,0),ff,a=.5,b=.2)
```

`optim` minimises `ff` w.r.t. the elements of `x`. It passes 0.5 and 0.2 to `ff` as the values for `a` and `b`. Of course, we are not restricted to passing simple constants: almost any R object could also be passed as an argument.

One irritation is worth being aware of.

```
ff <- function(res=1,...) res;f(r=2)
```

will return the answer 2 as a result of partial matching of argument names, even if you meant `r` to be part of the '...' argument. It is easy to be caught out by this. If you want '...' to be matched first, then it has to precede the arguments it might be confused with. So the following gives the answer 1:

```
ff <- function(...,res=1) res;f(r=2)
```

3.5 Useful built-in functions

The purpose of this chapter is to provide an introductory overview, not a reference, for R. So this section simply provides the information on where to locate the documentation for some useful standard built-in functions. R has an extensive help system, which can be accessed by typing `help.start()` at the command prompt, to obtain help in navigable HTML form, or by typing `?foo` at the command line, where `foo` is the function or other topic of interest.

Help topic	Subject covered
`?Arithmetic`	Standard arithmetic operators
`?Logic`	Standard logical operators
`?sqrt`	Square root and absolute value functions
`?Trig`	Trigonometric functions (`sin`, `cos`, etc.)
`?Hyperbolic`	Hyperbolic functions (`tanh`, etc.)
`?Special`	Special mathematical functions (Γ function, etc.)
`?pgamma`	Partial gamma function
`?Bessel`	Bessel functions
`?log`	Logarithmic functions
`?max`	Maximum, minimum and vectorised versions
`?round`	Rounding, truncating, etc.
`?distributions`	Statistical distributions built into R

The `?distributions` topic requires some more explanation. R has built-in functions for the `beta`, `binomial`, `cauchy`, `chisquared`, `exponential`, `f`, `gamma`, `geometric`, `hypergeometric`, `lnormal` (log-normal), `multinomial`, `nbinomial` (negative binomial), `normal`, `poisson`, `t`, `uniform` and `weibull` distributions. The R identifying names for these are shown in `courier` font in this list.

For each distribution, with name `dist`, say, there are four functions:

1. `ddist` is the probability (density) function of `dist`.
2. `pdist` is the cumulative distribution functions of `dist`.
3. `qdist` is the quantile function of `dist`.
4. `rdist` generates independent pseudorandom deviates from `dist`.

3.6 Object orientation and classes

Objects in R have classes, and R contains a mechanism by which different versions of a function may be used depending on an object's class. For example, the somewhat complicated list object returned from the linear modelling function `lm` has class `"lm"`:

```
> set.seed(0); n <- 100
> x <- runif(n); y <- x + rnorm(n)
> b <- lm(y~x)
> class(b)
[1] "lm"
```

This is why if we just type `b` or equivalently `print(b)` at the command prompt, then rather than getting a dull and lengthy printout of everything `b` contains (the default for a list), we get the much prettier result:

```
> print(b)

Call:
lm(formula = y ~ x)

Coefficients:
(Intercept)          x
   -0.05697    0.92746
```

What has happened here is that the `print` method function appropriate to the `"lm"` class has been invoked to do the printing. This function is called `print.lm` (type `stats:::print.lm` at the command prompt if you want to see what it looks like). The mechanism by which this happens involves method dispatch via a *generic* `print` function. If we examine the print function then it turns out to contain only a single line:

```
> print
function (x, ...)
UseMethod("print")
```

which tells R to use a function `print.foo` based on the class `foo` of the first argument `x`: there is a `print.default` to use if no `print.foo` method is found. Common uses of this sort of approach are for `print`, `summary` and `plot` functions.

At first sight it might appear that the existence of `print` methods is a nuisance if you want to know exactly what is inside an object, but actually it is no handicap. You can always call the default print method directly (e.g. `print.default(b)` prints out the contents of `b` in tedious and overwhelming detail). `str(b)` is usually a better bet, giving a summary of the structure of its argument. `names(b)` simply tells you the names of `b`'s elements.

Many object classes are quite closely related to other object classes. For example, generalised linear models share many features of linear models, and as a result there is much overlap in the structure of the objects returned by `lm` and `glm`. This overlap immediately raises the question of whether some `lm` methods could be used directly with `glm` objects without needing to be rewritten. The idea of *inheritance* of classes facilitates this. An object of one class can inherit from one or more other classes, and any method functions missing for this class can then default to the versions for the classes from which it inherits. For example,

```
> b1 <- glm(y~x)
> class(b1)
[1] "glm" "lm"
```

indicates that class `"glm"` inherits from class `"lm"`, which could also be tested using `inherits(b1,"lm")`. In the case of `b1` there is a `print.glm` method, but no `plot.glm`, so `plot(b1)` will actually use `plot.lm(b1)`.

We are free to add methods for existing generic functions and to create our own generics. As an example, the following code creates a version of the '`+`' operator that concatenates lines of poetry, which are given class `"poetry"`, and creates a `print` method for the class:

```
"+.poetry" <- function(a,b) {
 d <- paste(a,b,sep="\n")
 class(d) <- "poetry"
 d
}

print.poetry <- function(x) cat(x,"\n")
```

Note that `paste` is a function for pasting together character strings and `cat` is a basic text output function. Having provided these methods for objects of class `"poetry"`, here they are in action:

```
> a <- "twas brillig and the slithy toves"
> b <- "Did gyre and gimble in the wabe"
> d <- "All mimsy were the borogroves"
> class(a) <- class(b) <- class(d) <- "poetry"
> a + b + d
twas brillig and the slithy toves
Did gyre and gimble in the wabe
All mimsy were the borogroves
```

The mechanisms described here are a quite weak form of object orientation, known as `S3` classes and methods. A much fuller form of object orientation is provided by `S4` classes and methods in the R package `methods`. The `?Methods` help file contains very clear information on both approaches.

3.7 Conditional execution and loops

It is often necessary to evaluate different sets of R commands depending on whether some condition holds or not. When working with elements of vectors and arrays the efficient way to evaluate conditionally is to use the logical indexing methods of Section 3.3.4; otherwise R offers this structure:

```
if (condition) {
  statements 1
} else {
  statements 2
}
```

If the expression `condition` evaluates to `TRUE` then the expressions corresponding to `statements 1` are evaluated; otherwise `statements 2` are evaluated. The `else { ... }` part is optional: if it is omitted then nothing is evaluated when `condition` is `FALSE`, and R simply moves on to the next instruction. Nested `if` statements can be constructed in the obvious way (see also `?switch`):

```
if (condition 1) {
  statements 1
} else if (condition 2) {
  statements 2
} else {
  statement 3
}
```

Here is a simple example to simulate the tossing of a coin:

```
if (runif(1) > 0.5) cat("heads\n") else cat("tails\n")
```

Statements such as `a <-if (condition) foo else bar` are also commonly used.

Another essential programming task is looping. In R it is important to avoid looping over array elements whenever possible: the methods detailed in Section 3.3 are usually *much* more efficient. However, there are many legitimate uses of looping (where each iteration of the loop is doing lots of work), and R has five commands for loop implementation: `for`, `while`, `repeat`, `break` and `next`.

`for` is perhaps the most commonly used. It repeats a set of R commands once for each element of a vector. The syntax is

```
for (a in vec) {
  R code
}
```

where the `R code` in brackets is repeated for `a` set equal to each element of `vec` in turn. For example,

```
> vec <- c("I","am","bored")
> for (a in vec) cat(a," ")
I am bored
```

The commonest use of `for` is to loop over all integers between some limits. For example, `for (i in 1:10) {...}` evaluates the commands in `{...}` for $i = 1, 2, \ldots, 10$.

`while` executes a loop until a condition is no longer met. The syntax is

```
while (condition) {
  R code
}
```

So it repeatedly evaluates `R code` until `condition` no longer evaluates to `TRUE`. The following example iterates a simple ecological population model until some threshold population is exceeded:

```
N <- 2
while (N < 100) N <- N * exp(rnorm(1)*.1)
```

Notice that this is an example of a case where we cannot avoid looping by vectorizing: the computation is fundamentally iterative.

`break` and `next` are commands for modifying the looping behaviour from within the code being looped over. `break` causes an immediate exit from the loop. `next` causes the loop to skip directly to the next iteration.

The existence of `break` facilitates R's simplest looping instruction `repeat`, which simply repeats a set of instructions until a `break` is encountered. Here is a rewrite of the `while` population model example, using `repeat`. The logic is identical.

```
N <- 2
repeat {
  N <- N * exp(rnorm(1)*.1)
  if (N >= 100) break
}
```

Note the indentation of code within the loop: this is generally considered to be good practice because it improves readability.

3.7.1 Loop efficiency

The key to efficient programming in R is to make sure that each iteration of a loop is doing lots of work (with relatively little code). Looping over elements of arrays, and doing only a little work on each element, is usually very inefficient. To emphasise this, consider the example of multiplying two square matrices. The following R code compares the timings of a naive loop in R with use of '`%*%`' for the same task.

```
> n <- 100L
> A <- matrix(runif(n^2),n,n)
> B <- matrix(runif(n^2),n,n)
> C <- B*0
> system.time({
+ for (i in 1:n) for (j in 1:n) for (k in 1:n)
+ C[i,j] <- C[i,j] + A[i,k] * B[k,j]})
   user system elapsed
 11.213 0.012 11.223
> system.time(C <- A%*%B)
   user system elapsed
  0.004 0.000 0.002
```

The reason that the loop is so slow is that R is an interpreted language. At each iteration of the nested loop, `C[i,j] <-C[i,j] + A[i,k] *B[k,j]` has to be interpreted and evaluated, which takes far longer than the one addition and one multiplication that the expression actually results in. Of course, the naive loop is particularly boneheaded. We could much improve matters by replacing the inner loop with something vector oriented, thereby increasing the work done at each iteration while reducing the interpretation and evaluation overheads by a factor of around 100:

```
> system.time({
+ for (i in 1:n) for (j in 1:n)
```

```
+ C[i,j] <- sum(A[i,] * B[,j])})
   user  system elapsed
  0.224   0.000   0.223
```

This is better, but still around 50 times slower than the single line version.

When lengthy looping is unavoidable, as in truly iterative computations, then some improvements can be obtained by *byte-compilation* of R, using the standard package `compiler`. Here is an example of creating a byte-compiled function out of the second R loop using the `cmpfun` function:

```
> require(compiler)
> bad.loop <- cmpfun(function(A,B) {
+ C <- A*0
+ for (i in 1:n) for (j in 1:n)
+ C[i,j] <- sum(A[i,] * B[,j])
+ C
+ })
> system.time(C <- bad.loop(A,B))
   user  system elapsed
  0.108   0.000   0.108
```

A modest improvement, but still much slower than `A%*%B`.

3.8 Calling compiled code

There exist tasks for which R is simply inefficient, but this inefficiency is often easy to overcome by calling external compiled code from R. Interfaces exist for Fortran, C and C++, and it is possible to call back into R from compiled code. This section only considers the most basic interface for calling C code from R. Windows users will need the extra software provided at `http://cran.r-project.org/bin/windows/Rtools/`. Most systems based on some variety of Unix should already have the required tools available.

Consider the example of writing C code to implement the matrix multiplication loop from the previous section. Suppose that such a function is contained in `matmult.c`, as follows:

```
#include <math.h>
#include "matmult.h"

void matmult(double *A, double *B, double *C, int *n) {
  int i,j,k;
  for (i=0;i < *n;i++) for (j=0;j < *n;j++) {
    C[i + *n * j] = 0;for (k=0;k < *n;k++)
    C[i + *n * j] += A[i + *n * k] * B[k + *n * j];
  }
}
```

Notice the assumption that the matrices are stored columnwise in vectors, corresponding to R's underlying storage convention for matrices. There is a corresponding header file, `matmult.h`, containing

```
void matmult(double *A, double *B, double *C, int *n);
```

In the directory containing these files, the command line version of R can create a compiled version of this code, suitable for calling from R:

```
R CMD SHLIB matmult.c
```

will produce a shared object file `matmult.so` (`matmult.dll` on Windows). From within R we can now load this compiled object (either giving the full path to the file or using `setwd` to set the working directory to the directory containing it).

```
> dyn.load("matmult.so")
```

The routine itself can now be called, via R function `.C`:

```
> res <- .C("matmult",A=as.double(A),B=as.double(B),
+              C=as.double(C*0),n=as.integer(n))
> C <- matrix(res$C,n,n)
```

The arguments are explicitly converted to the type that the C code is expecting, using `as.double`, etc. For the matrices, what is passed to C is a pointer to the underlying vector of values (stored one column after another). `.C` actually copies all the objects that are its arguments, before passing pointers to these copies to the C code. `.C` then returns a list containing the copied arguments, with any modification of the copies that the C code has made.[9] For this example, the copy of `C` has been modified. In the above call, each argument has been given a name; for example, by using `A=as.double(A)` to name the first argument `A`. If names are omitted, then the elements of the return list are accessed by number in the usual way. Finally, notice that `res$C` had to be explicitly converted from a vector back to a matrix.

Applying `system.time` to the call to `matmult` reveals that it takes about three times as long as `A%*%B`. The speed penalty occurs because the given `C` code is itself inefficiently written: speeding it up by a factor of three is fairly easy, but also beyond the scope of this book. However, even this inefficient C code is much faster than anything we achieved using R loops for this task.

See `http://cran.r-project.org/doc/manuals/R-exts.html` for much more detail on this topic, but note that there are numerous

[9] Hence the C code does not modify the original R objects passed as arguments to `.C`

opportunities to call R routines from C if the header file `R.h` is included in the code. For example, `unif_rand()` and `norm_rand()` give access to R's uniform and standard normal pseudorandom number generators.

3.9 Good practice and debugging

R is enormously flexible: so flexible that it is easy to code in a very sloppy way that is likely to create hard-to-trace errors. For this reason it is worth imposing some self discipline when coding. Top of the list is to code for legibility. There are four obvious aspects to this:

1. Use meaningful object names, but try to keep them concise. For example, if coding up a model that you wrote down in terms of parameters α, β and γ, then refer to those parameters as `alpha`, `beta` and `gamma` in your code.
2. White space costs almost nothing, so use it to space out your expressions for readability. It is good practice to use white space around operators to avoid ambiguity. For example, `a <- 2` is unambiguous, whereas `a<-2` could be assignment or could be the logical result of testing if `a` is less than -2 (R would choose assignment, but code is clearer if you do not have to known that).
3. Use comments to explain code. Comments start with `#` and continue until a line end. Use them freely.
4. When coding a complex task, take care to structure your code carefully. Break the task down into functions, each performing a discrete, well-defined and comprehensible part of the overall work.

The second component of self-discipline is to resist the temptation to do everything interactively at the command prompt. R code for anything remotely complex should be coded up in a text file (which can be saved) and then pasted or `source`'d into R. Various R-based computing environments exist to make this way of working easier.

Finally, a word about debugging. Anyone writing remotely interesting code makes mistakes. Being very careful about writing down exactly what you want to do on paper before coding it can minimise the number of coding bugs, but will not eliminate them. It is good practice to assume that code is buggy until, after strenuous effort, you fail to find any more bugs. The temptation when confronted with a stubborn bug is to spend hours gazing at the code. This is usually a waste of time: if you were going to see the bug that way, you would probably have done so when first writing the code. It is more effective to apply the approach of scientific investigation to your code. Formulate hypotheses about what might be wrong (or even what should be right), and design experiments to test these hypotheses. When doing this it is helpful to print out intermediate values of computed quantities. However, to avoid wasting time, you should also learn how to use a debugger. In R the default debugging facilities are provided by `debug` and `trace` (see their help files for more information). However, I find the `mtrace` debugger from Mark Bravington's `debug` package to be much more useful.

You can also debug C (and other) code called from R using the GNU `gdb` debugger. For C/C++ on Linux, `nemiver` is somewhat easier to use than `gdb` (and in some respects more powerful). The `valgrind` memory error debugger can also be used for compiled code in R. Again, see `http://cran.r-project.org/doc/manuals/R-exts.html` for details.

Exercises

3.1 Computers do not represent most real numbers exactly. Rather, a real number is approximated by the nearest real number that can be represented exactly (*floating point number*), given some scheme for representing real numbers as fixed-length binary sequences. Often the approximation is not noticeable, but it can make a big difference relative to exact arithmetic (e.g., imagine that you want to know the difference between two distinct real numbers that are approximated by the *same* binary sequence). One consequence of working in *finite precision arithmetic* is that for any number x, there is a small number ϵ such that for all e, $|e| \leq |\epsilon|$, $x + e$ is indistinguishable from x.

a. Try out the following code to find the size of this number, when $x = 1$:

```
eps <- 1
x <- 1
while (x+eps != x) eps <- eps/2
eps/x
```

b. Confirm that the final `eps` here is close to the largest ϵ for which x and $x + \epsilon$ give rise to the same floating point number.

c. `2*eps` is stored in R as `.Machine$double.eps`. Confirm this.

d. Confirm that `eps/x` is the same for x = 1/8,1/4,1/2,1,2, 4 or 8.

e. Now try some numbers that are not exactly representable as modest powers of 2, and note the difference.

f. In terms of decimal digits, roughly how accurately are real numbers being represented here?

3.2 Rewrite the following to eliminate the loops, first using `apply` and then using `rowSums`:

```
X <- matrix(runif(100000),1000,100); z <- rep(0,1000)
for (i in 1:1000) {
  for (j in 1:100) z[i] <- z[i] + X[i,j]
}
```

Confirm that all three versions give the same answers, but that your rewrites are much faster than the original. (`system.time` is a useful function.)

3.3 Rewrite the following, replacing the loop with efficient code:

```
n <- 100000; z <- rnorm(n)
zneg <- 0;j <- 1
for (i in 1:n) {
  if (z[i]<0) {
    zneg[j] <- z[i]
    j <- j + 1
  }
}
```

Confirm that your rewrite is faster but gives the same result.

3.4 Run the following code:

```
set.seed(1); n <- 1000
A <- matrix(runif(n*n),n,n); x <- runif(n)
```

Evaluate $\mathbf{x}^{\mathrm{T}}\mathbf{A}\mathbf{x}$, $\text{tr}(\mathbf{A})$ and $\text{tr}(\mathbf{A}^{\mathrm{T}}\mathbf{W}\mathbf{A})$ where $\mathbf{W}$ is the diagonal matrix such that $W_{ii} = x_i$.

3.5 Consider solving the matrix equation $\mathbf{A}\mathbf{x} = \mathbf{y}$ for $\mathbf{x}$, where $\mathbf{y}$ is a known n vector and $\mathbf{A}$ is a known $n \times n$ matrix. The formal solution to the problem is $\mathbf{x} = \mathbf{A}^{-1}\mathbf{y}$, but it is possible to solve the equation directly, without actually forming $\mathbf{A}^{-1}$. This question explores this direct solution. Read the help file for `solve` before trying it.

a. First create an $\mathbf{A}$, $\mathbf{x}$ and $\mathbf{y}$ satisfying $\mathbf{A}\mathbf{x} = \mathbf{y}$.

```
set.seed(0); n <- 1000
A <- matrix(runif(n*n),n,n); x.true <- runif(n)
y <- A%*%x.true
```

The idea is to experiment with solving $\mathbf{A}\mathbf{x} = \mathbf{y}$ for $\mathbf{x}$, but with a known truth to compare the answer to.

b. Using `solve`, form the matrix $\mathbf{A}^{-1}$ explicitly and then form $\mathbf{x}_1 = \mathbf{A}^{-1}\mathbf{y}$. Note how long this takes. Also assess the mean absolute difference between `x1` and `x.true` (the approximate mean absolute 'error' in the solution).

c. Now use `solve` to directly solve for $\mathbf{x}$ without forming $\mathbf{A}^{-1}$. Note how long this takes and assess the mean absolute error of the result.

d. What do you conclude?

3.6 The empirical cumulative distribution function for a set of measurements $\{x_i : i = 1, \ldots n\}$ is

$$\hat{F}(x) = \frac{\#\{x_i < x\}}{n}$$

where $\#\{x_i < x\}$ denotes 'number of x_i values less than x'. When answering the following, try to ensure that your code is commented, clearly structured, and tested. To test your code, generate random samples using `rnorm`, `runif`, etc.

a. Write an R function that takes an unordered vector of observations `x` and returns the values of the empirical c.d.f. for each value, in the order corresponding to the original `x` vector. See `?sort.int`.
b. Modify your function to take an extra argument `plot.cdf`, that when `TRUE` will cause the empirical c.d.f. to be plotted as a step function over a suitable x range.

3.7 Try out the `debug` function on your function from the previous question. Then install the `debug` package from `CRAN` and see how using the `mtrace` function compares. To get started, take a look at the html help for the `debug` package.

3.8 In an R session containing nothing important, run the following code.

```
rm(list=ls())
hello2 <- function(name=NULL,n=3,dum=0) {
  txt <- paste(paste(rep("hello ",n),collapse=""),
                        name,"\n",sep="")
  cat(txt)
}
hello2(foo,2)
hello2("simon",2,foo)
```

Why does the first call to `hello2` generate an error, but not the second?

3.9 Work out the reasons for the differences and similarities in the results of calling `foo` and `bar` in the following code:

```
foo <- function() {
  print(parent.env(environment()))
  print(parent.frame())
}
bar <- function() foo()
foo()
bar()
```

4

Theory of maximum likelihood estimation

The use of maximum likelihood estimation rests on some general theory about log likelihoods and maximum likelihood estimates. This chapter briefly covers the derivation of the key results at a level suitable for ensuring their reliable application. See Cox and Hinkley (1974), Silvey (1970) and Davison (2003) for more detail.

4.1 Some properties of the expected log likelihood

Large sample theory for maximum likelihood estimators relies on some results for the expected log likelihood and on the observed likelihood tending to its expected value as the sample size tends to infinity. The results for the expected log likelihood are derived here. Recall that $l(\boldsymbol{\theta}) = \log f_{\theta}(\mathbf{y})$, and let $\boldsymbol{\theta}_t$ be the vector of true parameter values.

1.

$$E\left(\left.\frac{\partial l}{\partial \boldsymbol{\theta}}\right|_{\theta_t} \right) = \mathbf{0}, \tag{4.1}$$

where the expectation is taken at $\boldsymbol{\theta}_t$. Proof is straightforward provided that there is sufficient regularity to allow the order of differentiation and integration to be exchanged:

$$\begin{aligned} E\left\{ \frac{\partial}{\partial \boldsymbol{\theta}} \log f_{\theta}(\mathbf{y}) \right\} &= \int \frac{1}{f_{\theta}(\mathbf{y})} \frac{\partial f_{\theta}}{\partial \boldsymbol{\theta}} f_{\theta}(\mathbf{y}) d\mathbf{y} = \int \frac{\partial f_{\theta}}{\partial \boldsymbol{\theta}} d\mathbf{y} \\ &= \frac{\partial}{\partial \boldsymbol{\theta}} \int f_{\theta}(\mathbf{y}) d\mathbf{y} = \frac{\partial \mathbf{1}}{\partial \boldsymbol{\theta}} = \mathbf{0}. \end{aligned}$$

2.

$$\text{cov}\left(\left.\frac{\partial l}{\partial \boldsymbol{\theta}}\right|_{\theta_t} \right) = E\left(\left.\frac{\partial l}{\partial \boldsymbol{\theta}}\right|_{\theta_t} \left.\frac{\partial l}{\partial \boldsymbol{\theta}^{\mathrm{T}}}\right|_{\theta_t} \right), \tag{4.2}$$

which follows directly from the previous result and the definition of a covariance matrix, (1.4). Recall here that $\partial l / \partial \boldsymbol{\theta}$ is a column vector and $\partial l / \partial \boldsymbol{\theta}^{\mathrm{T}}$ a row vector.

3.

$$\mathcal{I} = E\left(\left.\frac{\partial l}{\partial \boldsymbol{\theta}}\right|_{\theta_t} \left.\frac{\partial l}{\partial \boldsymbol{\theta}^{\mathrm{T}}}\right|_{\theta_t} \right) = -E\left(\left.\frac{\partial^2 l}{\partial \boldsymbol{\theta} \partial \boldsymbol{\theta}^{\mathrm{T}}}\right|_{\theta_t} \right). \tag{4.3}$$

$\mathcal{I}$, is known as the *Fisher information matrix*. The terminology relates to the fact that a likelihood containing lots of information about $\boldsymbol{\theta}$ will be sharply peaked ($\mathcal{I}$ will have large magnitude eigenvalues), whereas a less informative likelihood will be less sharply peaked.

Proof is straightforward. From (4.1) we have

$$\int \frac{\partial \log f_\theta}{\partial \boldsymbol{\theta}} f_\theta(\mathbf{y}) d\mathbf{y} = \mathbf{0}$$
$$\Rightarrow \int \frac{\partial^2 \log f_\theta}{\partial \boldsymbol{\theta} \partial \boldsymbol{\theta}^{\mathrm{T}}} f_\theta(\mathbf{y}) + \frac{\partial \log f_\theta}{\partial \boldsymbol{\theta}} \frac{\partial f_\theta}{\partial \boldsymbol{\theta}^{\mathrm{T}}} d\mathbf{y} = \mathbf{0},$$

but $\dfrac{\partial \log f_\theta}{\partial \boldsymbol{\theta}^{\mathrm{T}}} = \dfrac{1}{f_\theta} \dfrac{\partial f_\theta}{\partial \boldsymbol{\theta}^{\mathrm{T}}}$, so

$$\int \frac{\partial^2 \log f_\theta}{\partial \boldsymbol{\theta} \partial \boldsymbol{\theta}^{\mathrm{T}}} f_\theta(\mathbf{y}) d\mathbf{y} = -\int \frac{\partial \log f_\theta}{\partial \boldsymbol{\theta}} \frac{\partial \log f_\theta}{\partial \boldsymbol{\theta}^{\mathrm{T}}} f_\theta(\mathbf{y}) d\mathbf{y},$$

and the result is proven.

4. The expected log likelihood has a global maximum at $\boldsymbol{\theta}_t$. i.e.

$$E\{l(\boldsymbol{\theta}_t)\} \geq E\{l(\boldsymbol{\theta})\} \ \forall \ \boldsymbol{\theta}. \tag{4.4}$$

Since log is a concave function, Jensen's inequality (1.10) implies that

$$E\left[\log\left\{ \frac{f_\theta(\mathbf{y})}{f_{\theta_t}(\mathbf{y})} \right\} \right] \leq \log\left[E\left\{ \frac{f_\theta(\mathbf{y})}{f_{\theta_t}(\mathbf{y})} \right\} \right]$$
$$= \log \int \frac{f_\theta(\mathbf{y})}{f_{\theta_t}(\mathbf{y})} f_{\theta_t}(\mathbf{y}) d\mathbf{y} = \log \int f_\theta(\mathbf{y}) d\mathbf{y} = \log(1) = 0,$$

and the result is proven.

5. The Cramér-Rao lower bound. $\mathcal{I}^{-1}$ provides a lower bound on the variance matrix of any unbiased estimator $\tilde{\boldsymbol{\theta}}$, in the sense that $\mathrm{cov}(\tilde{\boldsymbol{\theta}}) - \mathcal{I}^{-1}$ is positive semi-definite.

Proof: Since $f\partial \log f/\partial\theta = \partial f/\partial\theta$, $\partial\boldsymbol{\theta}_t/\partial\boldsymbol{\theta}_t^{\mathrm{T}} = \mathbf{I}$ and $\tilde{\boldsymbol{\theta}}$ is unbiased,

$$\int \tilde{\boldsymbol{\theta}} f_{\theta_t}(\mathbf{y})d\mathbf{y} = \boldsymbol{\theta}_t \Rightarrow \int \tilde{\boldsymbol{\theta}} \left.\frac{\partial \log f_{\theta_t}}{\partial \boldsymbol{\theta}_t^{\mathrm{T}}}\right|_{\theta_t} f_{\theta_t}(\mathbf{y})d\mathbf{y} = \mathbf{I}.$$

Hence, given (4.1), the matrix of covariances of elements of $\tilde{\boldsymbol{\theta}}_t$ with elements of $\partial \log f_{\theta_t}/\partial\boldsymbol{\theta}_t$ can be obtained:

$$\mathrm{cov}\left(\tilde{\boldsymbol{\theta}}, \left.\frac{\partial \log f_{\theta_t}}{\partial \boldsymbol{\theta}_t}\right|_{\theta_t}\right)$$
$$= E\left(\tilde{\boldsymbol{\theta}} \left.\frac{\partial \log f_{\theta_t}}{\partial \boldsymbol{\theta}_t^{\mathrm{T}}}\right|_{\theta_t}\right) - E(\tilde{\boldsymbol{\theta}})E\left(\left.\frac{\partial \log f_{\theta_t}}{\partial \boldsymbol{\theta}_t^{\mathrm{T}}}\right|_{\theta_t}\right) = \mathbf{I}.$$

Combining this with (4.2) we obtain the variance-covariance matrix,

$$\mathrm{cov}\begin{bmatrix} \tilde{\boldsymbol{\theta}} \\ \left.\frac{\partial \log f_{\theta_t}}{\partial \boldsymbol{\theta}_t}\right|_{\theta_t} \end{bmatrix} = \begin{bmatrix} \mathrm{cov}(\tilde{\boldsymbol{\theta}}) & \mathbf{I} \\ \mathbf{I} & \boldsymbol{\mathcal{I}} \end{bmatrix},$$

which is positive semi-definite by virtue of being a variance-covariance matrix. It follows that

$$\begin{bmatrix} \mathbf{I} & -\boldsymbol{\mathcal{I}}^{-1} \end{bmatrix} \begin{bmatrix} \mathrm{cov}(\tilde{\boldsymbol{\theta}}) & \mathbf{I} \\ \mathbf{I} & \boldsymbol{\mathcal{I}} \end{bmatrix} \begin{bmatrix} \mathbf{I} \\ -\boldsymbol{\mathcal{I}}^{-1} \end{bmatrix} = \mathrm{cov}(\tilde{\boldsymbol{\theta}}) - \boldsymbol{\mathcal{I}}^{-1}$$

is positive semi-definite, and the result is proven.

If the sense in which $\boldsymbol{\mathcal{I}}^{-1}$ is a lower bound is unclear, consider the variance of any linear transformation of the form $\mathbf{a}^{\mathrm{T}}\tilde{\boldsymbol{\theta}}$. By the result just proven, and the definition of positive semi-definiteness,

$$0 \le \mathbf{a}^{\mathrm{T}}\{\mathrm{cov}(\tilde{\boldsymbol{\theta}}) - \boldsymbol{\mathcal{I}}^{-1}\}\mathbf{a} = \mathrm{var}(\mathbf{a}^{\mathrm{T}}\tilde{\boldsymbol{\theta}}) - \mathbf{a}^{\mathrm{T}}\boldsymbol{\mathcal{I}}^{-1}\mathbf{a},$$

$\Rightarrow \mathrm{var}(\mathbf{a}^{\mathrm{T}}\tilde{\boldsymbol{\theta}}) \ge \mathbf{a}^{\mathrm{T}}\boldsymbol{\mathcal{I}}^{-1}\mathbf{a}$. For example, the lower bound on $\mathrm{var}(\tilde{\theta}_i)$ is given by the ith element on the leading diagonal of $\boldsymbol{\mathcal{I}}^{-1}$.

4.2 Consistency of MLE

Maximum likelihood estimators are usually consistent, meaning that as the sample size tends to infinity, $\hat{\boldsymbol{\theta}}$ tends to $\boldsymbol{\theta}_t$ (provided that the likelihood is informative about the parameters). This occurs because in regular situations $l(\boldsymbol{\theta})/n \to E\{l(\boldsymbol{\theta})\}/n$ as the sample size, n, tends to infinity, so that eventually the maximum of $l(\boldsymbol{\theta})$ and $E\{l(\boldsymbol{\theta})\}$ must coincide at $\boldsymbol{\theta}_t$ by (4.4). The result is easy to prove if the log likelihood can be broken down into a sum of independent components (usually one per observation), so that the law of large numbers implies convergence of the log likelihood to its expectation. Consistency can fail when the number of parameters is growing alongside the sample size in such a way that, for at least some parameters, the information per parameter is not increasing with sample size.

4.3 Large sample distribution of MLE

Taylor's theorem implies that

$$\left.\frac{\partial l}{\partial \boldsymbol{\theta}}\right|_{\hat{\theta}} \simeq \left.\frac{\partial l}{\partial \boldsymbol{\theta}}\right|_{\theta_t} + \left.\frac{\partial^2 l}{\partial \boldsymbol{\theta} \partial \boldsymbol{\theta}^{\mathrm{T}}}\right|_{\theta_t} (\hat{\boldsymbol{\theta}} - \boldsymbol{\theta}_t)$$

with equality in the large sample limit, for which $\hat{\boldsymbol{\theta}} - \boldsymbol{\theta}_t \to 0$. From the definition of $\hat{\boldsymbol{\theta}}$, the left-hand side is $\mathbf{0}$. So assuming $\mathcal{I}/n$ is constant (at least in the $n \to \infty$ limit), then as the sample size tends to infinity,

$$\frac{1}{n}\left.\frac{\partial^2 l}{\partial \boldsymbol{\theta} \partial \boldsymbol{\theta}^{\mathrm{T}}}\right|_{\theta_t} \to -\frac{\mathcal{I}}{n}, \quad \text{while} \quad \left.\frac{\partial l}{\partial \boldsymbol{\theta}}\right|_{\theta_t}$$

is a random vector with mean $\mathbf{0}$ and covariance matrix $\mathcal{I}$ by (4.2) and (4.1).[1] Therefore in the large sample limit,

$$\hat{\boldsymbol{\theta}} - \boldsymbol{\theta}_t \sim \mathcal{I}^{-1} \left.\frac{\partial l}{\partial \boldsymbol{\theta}}\right|_{\theta_t},$$

implying that $E(\hat{\boldsymbol{\theta}} - \boldsymbol{\theta}_t) = 0$ and var $(\hat{\boldsymbol{\theta}} - \boldsymbol{\theta}_t) = \mathcal{I}^{-1}$. Hence in regular situations in the large sample limit, maximum likelihood estimators are unbiased and achieve the Cramér-Rao lower bound. This partly accounts for their popularity.

It remains to establish the large sample distribution of $\hat{\boldsymbol{\theta}} - \boldsymbol{\theta}_t$. In the case in which the likelihood is based on independent observations, then $l(\boldsymbol{\theta}) = \sum_i l_i(\boldsymbol{\theta})$, where l_i denotes the contribution to the log likelihood from the ith observation. In that case $\partial l / \partial \boldsymbol{\theta} = \sum_i \partial l_i / \partial \boldsymbol{\theta}$, so that $\partial l / \partial \boldsymbol{\theta}$ is a sum of independent random variables. Hence, under mild conditions, the central limit theorem applies, and in the large sample limit

$$\hat{\boldsymbol{\theta}} \sim N(\boldsymbol{\theta}_t, \mathcal{I}^{-1}). \tag{4.5}$$

[1] In the limit the random deviation of $n^{-1}\partial^2 l / \partial \boldsymbol{\theta} \partial \boldsymbol{\theta}^{\mathrm{T}}$ from its expected value, $n^{-1}\mathcal{I}$, is negligible relative to $n^{-1}\mathcal{I}$ itself (provided it is positive definite). This is never the case for the $\partial l / \partial \boldsymbol{\theta}$, because its expected value is zero.

In any circumstance in which (4.5) holds, it is also valid to use $-\partial^2 l/\partial\boldsymbol{\theta}\partial\boldsymbol{\theta}^{\mathrm{T}}$ in place of $\boldsymbol{\mathcal{I}}$ itself. When the likelihood is not based on independent observations, it is very often the case that $\partial l/\partial\boldsymbol{\theta}$ has a limiting normal distribution so that (4.5) holds anyway. The key is that the expected information increases without limit with increasing sample size. In any case, achievement of the Cramér-Rao lower bound does not depend on normality.

4.4 Distribution of the generalised likelihood ratio statistic

Consider testing:

$$H_0 : \mathbf{R}(\boldsymbol{\theta}) = \mathbf{0} \quad \text{vs.} \quad H_1 : \mathbf{R}(\boldsymbol{\theta}) \neq \mathbf{0},$$

where $\mathbf{R}$ is a vector-valued function of $\boldsymbol{\theta}$, such that H_0 imposes r restrictions on the parameter vector. If H_0 is true, then in the limit as $n \to \infty$,

$$2\lambda = 2\{l(\hat{\boldsymbol{\theta}}) - l(\hat{\boldsymbol{\theta}}_0)\} \sim \chi^2_r, \tag{4.6}$$

where l is the log-likelihood function and $\hat{\boldsymbol{\theta}}$ is the MLE of $\boldsymbol{\theta}$. $\hat{\boldsymbol{\theta}}_0$ is the value of $\boldsymbol{\theta}$ maximising the likelihood subject to the constraint $\mathbf{R}(\boldsymbol{\theta}) = \mathbf{0}$. This result is used to calculate approximate p-values for the test.

To derive (4.6), first re-parameterise so that $\boldsymbol{\theta}^{\mathrm{T}} = (\boldsymbol{\psi}^{\mathrm{T}}, \boldsymbol{\gamma}^{\mathrm{T}})$, where $\boldsymbol{\psi}$ is r dimensional and the null hypothesis can be rewritten $H_0 : \boldsymbol{\psi} = \boldsymbol{\psi}_0$. Such re-parameterisation is always possible, but is only necessary for deriving (4.6), not for its use.

Let the unrestricted MLE be $(\hat{\boldsymbol{\psi}}^{\mathrm{T}}, \hat{\boldsymbol{\gamma}}^{\mathrm{T}})$, and let $(\boldsymbol{\psi}_0^{\mathrm{T}}, \hat{\boldsymbol{\gamma}}_0^{\mathrm{T}})$ be the MLE under the restrictions defining the null hypothesis. To make progress, $\hat{\boldsymbol{\gamma}}_0$ must be expressed in terms of $\hat{\boldsymbol{\psi}}$, $\hat{\boldsymbol{\gamma}}$ and $\boldsymbol{\psi}_0$. Taking a Taylor expansion of l around the unrestricted MLE, $\hat{\boldsymbol{\theta}}$, yields

$$l(\boldsymbol{\theta}) \simeq l(\hat{\boldsymbol{\theta}}) - \frac{1}{2}\left(\boldsymbol{\theta} - \hat{\boldsymbol{\theta}}\right)^{\mathrm{T}} \mathbf{H} \left(\boldsymbol{\theta} - \hat{\boldsymbol{\theta}}\right), \tag{4.7}$$

where $H_{i,j} = -\left.\partial^2 l/\partial\theta_i\partial\theta_j\right|_{\hat{\boldsymbol{\theta}}}$. Exponentiating produces

$$L(\boldsymbol{\theta}) \simeq L(\hat{\boldsymbol{\theta}}) \exp\left[-\left(\boldsymbol{\theta} - \hat{\boldsymbol{\theta}}\right)^{\mathrm{T}} \mathbf{H} \left(\boldsymbol{\theta} - \hat{\boldsymbol{\theta}}\right)/2\right]$$

(i.e. the likelihood can be approximated by a function proportional to the p.d.f. of an $N(\hat{\boldsymbol{\theta}}, \mathbf{H}^{-1})$ random vector). So, in the large sample limit and defining $\boldsymbol{\Sigma} = \mathbf{H}^{-1}$, the likelihood is proportional to the p.d.f. of

$$N\left(\left[\begin{array}{c}\hat{\boldsymbol{\psi}}\\ \hat{\boldsymbol{\gamma}}\end{array}\right], \left[\begin{array}{cc}\boldsymbol{\Sigma}_{\psi\psi} & \boldsymbol{\Sigma}_{\psi\gamma}\\ \boldsymbol{\Sigma}_{\gamma\psi} & \boldsymbol{\Sigma}_{\gamma\gamma}\end{array}\right]\right).$$

If $\boldsymbol{\psi} = \boldsymbol{\psi}_0$ then this p.d.f. will be maximised by[2] $\hat{\boldsymbol{\gamma}}_0 = E(\boldsymbol{\gamma}|\boldsymbol{\psi}_0)$, which, from the results of Section 1.6.3, is

$$\hat{\boldsymbol{\gamma}}_0 = \hat{\boldsymbol{\gamma}} + \boldsymbol{\Sigma}_{\gamma\psi}\boldsymbol{\Sigma}_{\psi\psi}^{-1}(\boldsymbol{\psi}_0 - \hat{\boldsymbol{\psi}}). \tag{4.8}$$

If the null hypothesis is true, then in the large sample limit $\hat{\boldsymbol{\psi}} \to \boldsymbol{\psi}_0$ (in probability) so that the approximate likelihood tends to the true likelihood, and we can expect (4.8) to hold for the maximisers of the true likelihood.

It helps to express (4.8) in terms of the partitioned version of $\mathbf{H}$. Writing $\boldsymbol{\Sigma}\mathbf{H} = \mathbf{I}$ in partitioned form

$$\left[\begin{array}{cc}\boldsymbol{\Sigma}_{\psi\psi} & \boldsymbol{\Sigma}_{\psi\gamma}\\ \boldsymbol{\Sigma}_{\gamma\psi} & \boldsymbol{\Sigma}_{\gamma\gamma}\end{array}\right]\left[\begin{array}{cc}\mathbf{H}_{\psi\psi} & \mathbf{H}_{\psi\gamma}\\ \mathbf{H}_{\gamma\psi} & \mathbf{H}_{\gamma\gamma}\end{array}\right] = \left[\begin{array}{cc}\mathbf{I} & \mathbf{0}\\ \mathbf{0} & \mathbf{I}\end{array}\right],$$

and multiplying out, results in two useful equations:

$$\boldsymbol{\Sigma}_{\psi\psi}\mathbf{H}_{\psi\psi} + \boldsymbol{\Sigma}_{\psi\gamma}\mathbf{H}_{\gamma\psi} = \mathbf{I} \text{ and } \boldsymbol{\Sigma}_{\psi\psi}\mathbf{H}_{\psi\gamma} + \boldsymbol{\Sigma}_{\psi\gamma}\mathbf{H}_{\gamma\gamma} = \mathbf{0}. \tag{4.9}$$

Rearranging (4.9) while noting that, by symmetry, $\mathbf{H}_{\psi\gamma}^{\mathrm{T}} = \mathbf{H}_{\gamma\psi}$ and $\boldsymbol{\Sigma}_{\psi\gamma}^{\mathrm{T}} = \boldsymbol{\Sigma}_{\gamma\psi}$, yields

$$\boldsymbol{\Sigma}_{\psi\psi}^{-1} = \mathbf{H}_{\psi\psi} - \mathbf{H}_{\psi\gamma}\mathbf{H}_{\gamma\gamma}^{-1}\mathbf{H}_{\gamma\psi} \tag{4.10}$$

and $-\mathbf{H}_{\gamma\gamma}^{-1}\mathbf{H}_{\gamma\psi} = \boldsymbol{\Sigma}_{\gamma\psi}\boldsymbol{\Sigma}_{\psi\psi}^{-1}$. Substituting the latter into (4.8), we obtain

$$\hat{\boldsymbol{\gamma}}_0 = \hat{\boldsymbol{\gamma}} + \mathbf{H}_{\gamma\gamma}^{-1}\mathbf{H}_{\gamma\psi}(\hat{\boldsymbol{\psi}} - \boldsymbol{\psi}_0). \tag{4.11}$$

Now provided that the null hypothesis is true so that $\hat{\boldsymbol{\psi}}$ is close to $\boldsymbol{\psi}_0$, we can reuse the expansion (4.7) and write the log likelihood at the restricted MLE as

$$l(\boldsymbol{\psi}_0, \hat{\boldsymbol{\gamma}}_0) \simeq l(\hat{\boldsymbol{\psi}}, \hat{\boldsymbol{\gamma}}) - \frac{1}{2}\left[\begin{array}{c}\boldsymbol{\psi}_0 - \hat{\boldsymbol{\psi}}\\ \hat{\boldsymbol{\gamma}}_0 - \hat{\boldsymbol{\gamma}}\end{array}\right]^{\mathrm{T}}\mathbf{H}\left[\begin{array}{c}\boldsymbol{\psi}_0 - \hat{\boldsymbol{\psi}}\\ \hat{\boldsymbol{\gamma}}_0 - \hat{\boldsymbol{\gamma}}\end{array}\right].$$

Hence

$$2\lambda = 2\{l(\hat{\boldsymbol{\psi}}, \hat{\boldsymbol{\gamma}}) - l(\boldsymbol{\psi}_0, \hat{\boldsymbol{\gamma}}_0)\} \simeq \left[\begin{array}{c}\boldsymbol{\psi}_0 - \hat{\boldsymbol{\psi}}\\ \hat{\boldsymbol{\gamma}}_0 - \hat{\boldsymbol{\gamma}}\end{array}\right]^{\mathrm{T}}\mathbf{H}\left[\begin{array}{c}\boldsymbol{\psi}_0 - \hat{\boldsymbol{\psi}}\\ \hat{\boldsymbol{\gamma}}_0 - \hat{\boldsymbol{\gamma}}\end{array}\right].$$

[2] See Section 1.4.2 and Figure 1.3 if this is unclear.

Substituting for $\hat{\gamma}_0$ from (4.11) and writing out $\mathbf{H}$ in partitioned form gives

$$\begin{aligned} 2\lambda &\simeq \begin{bmatrix} \psi_0 - \hat{\psi} \\ \mathbf{H}_{\gamma\gamma}^{-1}\mathbf{H}_{\gamma\psi}(\hat{\psi} - \psi_0) \end{bmatrix}^{\mathrm{T}} \begin{bmatrix} \mathbf{H}_{\psi\psi} & \mathbf{H}_{\psi\gamma} \\ \mathbf{H}_{\gamma\psi} & \mathbf{H}_{\gamma\gamma} \end{bmatrix} \begin{bmatrix} \psi_0 - \hat{\psi} \\ \mathbf{H}_{\gamma\gamma}^{-1}\mathbf{H}_{\gamma\psi}(\hat{\psi} - \psi_0) \end{bmatrix} \\ &= (\hat{\psi} - \psi_0)^{\mathrm{T}} \left[\mathbf{H}_{\psi\psi} - \mathbf{H}_{\psi\gamma}\mathbf{H}_{\gamma\gamma}^{-1}\mathbf{H}_{\gamma\psi}\right](\hat{\psi} - \psi_0) \\ &= (\hat{\psi} - \psi_0)^{\mathrm{T}}\boldsymbol{\Sigma}_{\psi\psi}^{-1}(\hat{\psi} - \psi_0), \end{aligned}$$

where the final equality follows from (4.10). If H_0 is true, then as $n \to \infty$ this expression will tend towards exactness as $\hat{\psi} \to \psi_0$. Furthermore, provided $\mathbf{H} \to \mathcal{I}$ as $n \to \infty$, then $\boldsymbol{\Sigma}$ tends to $\mathcal{I}^{-1}$, and hence $\boldsymbol{\Sigma}_{\psi\psi}$ tends to the covariance matrix of $\hat{\psi}$, by (4.5). Hence, by the asymptotic normality of the MLE $\hat{\psi}$, $2\lambda \sim \chi^2_r$, under H_0.

4.5 Regularity conditions

The preceding results depend on some assumptions.

1. The densities defined by distinct values of $\boldsymbol{\theta}$ are distinct. If this is not the case the parameters need not be *identifiable*, and there is no guarantee of consistency.
2. $\boldsymbol{\theta}_t$ is interior to the space of possible parameter values. This is necessary in order to be able to approximate the log likelihood by a Taylor expansion in the vicinity of $\boldsymbol{\theta}_t$.
3. Within some neighbourhood of $\boldsymbol{\theta}_t$, the first three derivatives of the log likelihood exist and are bounded, while the Fisher information matrix satisfies (4.3) and is positive definite and finite. The various Taylor expansions and the arguments leading to (4.5) depend on this.

When these assumptions are met the results of this section are very general and apply in many situations well beyond the i.i.d setting. When they are not met, some or all of the results of Sections 4.2 to 4.4 will fail.

4.6 AIC: Akaike's information criterion

As briefly introduced in Section 2.4.5, an appealing approach to model selection is to select the model that appears to be as close to the truth as possible, in the Kullback-Leibler sense of minimising

$$K(f_\theta, f_t) = \int \{\log f_t(\mathbf{y}) - \log f_\theta(\mathbf{y})\} f_t(\mathbf{y})d\mathbf{y}, \tag{4.12}$$

where f_t is the true density of $\mathbf{y}$ and f_θ is the model approximation to it. To make this aspiration practical, we need to choose some version of K

that can be estimated, and it turns out that the expected value of $K(f_{\hat{\theta}}, f_t)$ is tractable, where $\hat{\boldsymbol{\theta}}$ is the MLE.

Although we cannot compute it, consider the value of $\boldsymbol{\theta}$ that would minimise (4.12), and denote it by $\boldsymbol{\theta}_K$. Now consider the Taylor expansion

$$\log f_{\hat{\theta}}(\mathbf{y}) \simeq \log f_{\theta_K}(\mathbf{y}) + (\hat{\boldsymbol{\theta}} - \boldsymbol{\theta}_K)^{\mathrm{T}} \left.\frac{\partial \log f_\theta}{\partial \boldsymbol{\theta}}\right|_{\theta_K} + \frac{1}{2}(\hat{\boldsymbol{\theta}} - \boldsymbol{\theta}_K)^{\mathrm{T}} \left.\frac{\partial^2 \log f_\theta}{\partial \boldsymbol{\theta} \partial \boldsymbol{\theta}^{\mathrm{T}}}\right|_{\theta_K} (\hat{\boldsymbol{\theta}} - \boldsymbol{\theta}_K). \quad (4.13)$$

If $\boldsymbol{\theta}_K$ minimises K then $\int \partial \log f_\theta / \partial \theta |_{\theta_K} f_t d\mathbf{y} = 0$, so substituting (4.13) into $K(f_{\hat{\theta}}, f_t)$, while *treating* $\hat{\boldsymbol{\theta}}$ *as fixed*[3] results in

$$K(f_{\hat{\theta}}, f_t) \simeq K(f_{\theta_K}, f_t) + \frac{1}{2}(\hat{\boldsymbol{\theta}} - \boldsymbol{\theta}_K)^{\mathrm{T}} \boldsymbol{\mathcal{I}}_{\theta_K} (\hat{\boldsymbol{\theta}} - \boldsymbol{\theta}_K), \quad (4.14)$$

where $\boldsymbol{\mathcal{I}}_{\theta_K}$ is the information matrix at $\boldsymbol{\theta}_K$. Now assume that the model is sufficiently correct that $E(\hat{\boldsymbol{\theta}}) \simeq \boldsymbol{\theta}_K$ and $\text{cov}(\hat{\boldsymbol{\theta}}) \simeq \boldsymbol{\mathcal{I}}_{\theta_K}$, at least for large samples. In this case, and reusing results from the end of Section 4.4,

$$E\{l(\hat{\boldsymbol{\theta}}) - l(\boldsymbol{\theta}_K)\} \simeq E\left\{\frac{1}{2}(\hat{\boldsymbol{\theta}} - \boldsymbol{\theta}_K)^{\mathrm{T}} \boldsymbol{\mathcal{I}}_{\theta_K} (\hat{\boldsymbol{\theta}} - \boldsymbol{\theta}_K)\right\} \simeq p/2 \quad (4.15)$$

where p is the dimension of $\boldsymbol{\theta}$. So taking expectations of (4.14) and substituting an approximation from (4.15),

$$EK(f_{\hat{\theta}}, f_t) \simeq K(f_{\theta_K}, f_t) + p/2. \quad (4.16)$$

Since this still involves the unknownable f_t, consider

$$\begin{aligned} E\{-l(\hat{\boldsymbol{\theta}})\} &= E[-l(\boldsymbol{\theta}_K) - \{l(\hat{\boldsymbol{\theta}}) - l(\boldsymbol{\theta}_K)\}] \\ &\simeq -\int \log\{f_{\theta_K}(\mathbf{y})\} f_t(\mathbf{y}) d\mathbf{y} - p/2 \text{ by (4.15)} \\ &= K(f_{\theta_K}, f_t) - p/2 - \int \log\{f_t(\mathbf{y})\} f_t(\mathbf{y}) d\mathbf{y}. \end{aligned}$$

Using this result to eliminate $K(f_{\theta_K}, f_t)$ from (4.16) suggests the estimate

$$\widehat{EK(f_{\hat{\theta}}, f_t)} = -l(\hat{\boldsymbol{\theta}}) + p + \int \log\{f_t(\mathbf{y})\} f_t(\mathbf{y}) d\mathbf{y}.$$

[3] By treating $\hat{\boldsymbol{\theta}}$ as fixed we are effectively assessing the expected likelihood ratio between model and truth for new data.

Since the last term on the right-hand side only involves the truth, this last estimate is minimised by whichever model minimises

$$\mathrm{AIC} = -2l(\hat{\boldsymbol{\theta}}) + 2p,$$

where the factor of 2 is by convention, to put AIC on the same scale as 2λ from Section 4.4. A possible concern here is that (4.15) is not justified if the model is oversimplified and hence poor, but in practice this is unproblematic, because the log likelihood decreases sharply as the approximation deteriorates. See Davison (2003) for a fuller derivation.

An objection to AIC is that it is not consistent: as $n \to \infty$, the probability of selecting the correct model does not tend to 1. For nested models (4.6) states that the difference in $-2l(\hat{\theta})$ between the true model and an overly complex model follows a χ^2_r distribution, where r is the number of spurious parameters. Neither χ^2_r nor $2p$ depends on n, so the probability of selecting the overly complex model by AIC is nonzero and independent of n (for n large). The same objection could also be made about hypothesis testing, unless we allow the accept/reject threshold to change with n.[4]

Exercises

4.1 The double exponential distribution has p.d.f. $f(x) = e^{-|x-\mu|/\sigma}/(2\sigma)$ where μ and σ are parameters. Obtain maximum likelihood estimates of μ and σ, given observations $x_1, x_2, \ldots, x_n$. (assume that n is even, the x_i are unique and $x_i \neq \mu$). Comment on the uniqueness of your estimates.

4.2 A random variable X has p.d.f. $f(x) = (b-a)^{-1}$ if $a \leq x \leq b$ and 0 otherwise. Given observations $x_1, x_2, \ldots, x_n$, find the maximum likelihood estimates of a and b. Are the corresponding estimators unbiased? Why is (4.5) inapplicable in this case?

4.3 Random variables X and Y have joint p.d.f. $f(x, y) = kx^{\alpha}y^{\beta}$ $0 \leq x \leq 1,\ 0 \leq y \leq 1$. Assume that you have n independent pairs of observations (x_i, y_i). (a) Evaluate k in terms of the α and β. (b) Find the maximum likelihood estimators of α and β. (c) Find approximate variances of $\hat{\alpha}$ and $\hat{\beta}$

[4] The inconsistency of AIC is not in itself the reason for the empirical observation that AIC tends to select increasingly complex models as n increases. If the true model is among those considered then (4.6) does not imply that the probability of rejecting it increases with sample size. However, if all the models under consideration are wrong, then we will tend to select increasingly complex approximations as the sample size increases and the predictive disadvantages of complexity diminish.

4.4 Suppose that you have n independent measurements of times between major aircraft disasters, t_i, and believe that the probability density function for the t_i's is of the form: $f(t) = ke^{-\lambda t^2}$ $t \geq 0$ where λ and k are the same for all i. (a) By considering the normal p.d.f., show that $k = \sqrt{4\lambda/\pi}$. (b) Obtain a maximum likelihood estimator for λ. (c) Given observations of T_i (in days) of: 243, 14, 121, 63, 45, 407 and 34 use a generalised likelihood ratio test to test $H_0 : \lambda = 10^{-4}$ against the alternative of no restriction on λ at the 5% significance level. Note that if $V \sim \chi_1^2$ then $\Pr[V \leq 3.841] = 0.95$.

5

Numerical maximum likelihood estimation

The theory of maximum likelihood estimation provides very general tools for inference using statistical models, provided we can evaluate the log likelihood and its first two derivatives and maximise the likelihood with respect to its parameters. For most interesting models we can not do this entirely analytically, and must use numerical methods for parts of the enterprise. The second-order Taylor expansion is again pivotal.[1]

5.1 Numerical optimisation

Most optimisation literature and software, including in R, concentrates on the minimisation of functions. This section follows this convention, bearing in mind that our goal of maximising log likelihoods can always be achieved by minimising negative log likelihoods. Generically, then, we are interested in automatic methods for finding

$$\hat{\boldsymbol{\theta}} = \underset{\theta}{\operatorname{argmin}} f(\boldsymbol{\theta}). \tag{5.1}$$

There are some very difficult problems in this class, so some restrictions are needed. Specifically, assume that the *objective function*, f, is a sufficiently smooth function, bounded below, and that the elements of $\boldsymbol{\theta}$ are unrestricted real parameters. So f might be a negative log likelihood, for example. f may also depend on other known parameters and data, but there is no need to clutter up the notation with these. The assumption that $\boldsymbol{\theta}$ is unrestricted means that, if we want to put restrictions on $\boldsymbol{\theta}$, we need to be able to implement them by writing $\boldsymbol{\theta} = \mathbf{r}(\boldsymbol{\theta}_r)$, where $\mathbf{r}$ is a known function and $\boldsymbol{\theta}_r$ is a set of unrestricted parameters. Then the problem becomes $\min_{\theta_r} f\{\mathbf{r}(\boldsymbol{\theta}_r)\}$.

[1] Some numerical matrix algebra is also taken for granted here, but Appendix B introduces most of what is needed.

Even given these assumptions, it is not possible to guarantee finding a solution to (5.1) unless we know that f is convex, which is generally an assumption too far. Pragmatically, the best we can hope for is to develop methods to find a *local minimum*; that is, a point $\hat{\boldsymbol{\theta}}$ such that $f(\hat{\boldsymbol{\theta}} + \boldsymbol{\Delta}) \geq f(\hat{\boldsymbol{\theta}})$, for any *sufficiently small* perturbation $\boldsymbol{\Delta}$. The resulting methods are adequate for many statistical problems.

5.1.1 Newton's method

A very successful optimisation method is based on iteratively approximating f by a truncated Taylor expansion and seeking the minimum of the approximation at each step. With a little care, this can be made into a method that is guaranteed[2] to converge to a local minimum. Taylor's theorem states that if f is a twice continuously differentiable function of $\boldsymbol{\theta}$, and $\boldsymbol{\Delta}$ is of the same dimension as $\boldsymbol{\theta}$, then for some $t \in (0, 1)$,

$$f(\boldsymbol{\theta} + \boldsymbol{\Delta}) = f(\boldsymbol{\theta}) + \nabla f(\boldsymbol{\theta})^{\mathrm{T}}\boldsymbol{\Delta} + \frac{1}{2}\boldsymbol{\Delta}^{\mathrm{T}}\nabla^2 f(\boldsymbol{\theta} + t\boldsymbol{\Delta})\boldsymbol{\Delta} \qquad (5.2)$$

$$\text{where } \nabla f(\boldsymbol{\theta}^*) = \left.\frac{\partial f}{\partial \boldsymbol{\theta}}\right|_{\theta^*} \text{ and } \nabla^2 f(\boldsymbol{\theta}^*) = \left.\frac{\partial^2 f}{\partial \boldsymbol{\theta} \partial \boldsymbol{\theta}^{\mathrm{T}}}\right|_{\theta^*}.$$

From (5.2), the condition $f(\hat{\boldsymbol{\theta}} + \boldsymbol{\Delta}) \geq f(\hat{\boldsymbol{\theta}})$, for any *sufficiently small* perturbation $\boldsymbol{\Delta}$, is equivalent to

$$\nabla f(\hat{\boldsymbol{\theta}}) = \mathbf{0} \text{ and } \nabla^2 f(\hat{\boldsymbol{\theta}}) \text{ positive semi-definite,} \qquad (5.3)$$

which are the useful conditions for a minimum.

A second consequence of (5.2) is that for sufficiently small α, a $\boldsymbol{\theta}$ that is not a turning point, and *any* positive definite matrix $\mathbf{H}$ of appropriate dimension, then $f\{\boldsymbol{\theta} - \alpha\mathbf{H}\nabla f(\boldsymbol{\theta})\} < f(\boldsymbol{\theta})$. Under the given conditions we can approximate f by a first-order Taylor approximation. Hence, in the small α limit, $f\{\boldsymbol{\theta} - \alpha\mathbf{H}\nabla f(\boldsymbol{\theta})\} = f(\boldsymbol{\theta}) - \alpha\nabla f(\boldsymbol{\theta})^{\mathrm{T}}\mathbf{H}\nabla f(\boldsymbol{\theta}) < f(\boldsymbol{\theta})$, where the inequality follows from the fact that $\nabla f(\boldsymbol{\theta})^{\mathrm{T}}\mathbf{H}\nabla f(\boldsymbol{\theta}) > 0$ since $\mathbf{H}$ is positive definite and $\nabla f(\boldsymbol{\theta}) \neq \mathbf{0}$. In short,

$$\Delta = -\mathbf{H}\nabla f(\boldsymbol{\theta}) \qquad (5.4)$$

[2] The statistical literature contains many statements about the possibility of Newton's method diverging, and the consequent difficulty of guaranteeing convergence. These statements are usually outdated, as a look at any decent textbook on optimisation shows.

is a *descent direction* if $\mathbf{H}$ is any positive definite matrix. After Taylor's theorem, this is probably the second most important fact in the optimisation of smooth functions.

Now consider Newton's method itself. Suppose that we have a guess, $\boldsymbol{\theta}'$, at the parameters minimising $f(\boldsymbol{\theta})$. Taylor's theorem implies that

$$f(\boldsymbol{\theta}' + \boldsymbol{\Delta}) \simeq f(\boldsymbol{\theta}') + \nabla f(\boldsymbol{\theta}')^{\mathrm{T}}\boldsymbol{\Delta} + \frac{1}{2}\boldsymbol{\Delta}^{\mathrm{T}}\nabla^2 f(\boldsymbol{\theta}')\boldsymbol{\Delta}.$$

Provided that $\nabla^2 f(\boldsymbol{\theta}')$ is positive semi-definite, the right-hand-side of this expression can be minimised by differentiating with respect to $\boldsymbol{\Delta}$ and setting the result to zero, which implies that

$$\nabla^2 f(\boldsymbol{\theta}')\boldsymbol{\Delta} = -\nabla f(\boldsymbol{\theta}'). \tag{5.5}$$

So, in principle, we simply solve for $\boldsymbol{\Delta}$ given $\boldsymbol{\theta}'$ and update $\boldsymbol{\theta}' \leftarrow \boldsymbol{\theta}' + \boldsymbol{\Delta}$ repeatedly until the conditions (5.3) are met. By Taylor's theorem itself, this process must converge if we start out close enough to the minimising $\hat{\boldsymbol{\theta}}$. But if we knew how to do that we perhaps would not need to be using Newton's method in the first place.

The method should converge when started from parameter guesses that are a long way from $\hat{\boldsymbol{\theta}}$, requiring two modifications of the basic iteration:

1. $\nabla^2 f(\boldsymbol{\theta}')$ is only guaranteed to be positive (semi) definite close to $\hat{\boldsymbol{\theta}}$. So $\nabla^2 f(\boldsymbol{\theta}')$ must be modified to make it positive definite, if it is not. The obvious alternatives are (i) to replace $\nabla^2 f(\boldsymbol{\theta}')$ by $\nabla^2 f(\boldsymbol{\theta}') + \delta\mathbf{I}$, where δ is chosen to be just large enough to achieve positive definiteness;[3] or (ii) take the symmetric eigen-decomposition $\nabla^2 f(\boldsymbol{\theta}') = \mathbf{U}\boldsymbol{\Lambda}\mathbf{U}^{\mathrm{T}}$, where $\boldsymbol{\Lambda}$ is the diagonal matrix of eigenvalues, and replace $\nabla^2 f(\boldsymbol{\theta}')$ by $\mathbf{U}\tilde{\boldsymbol{\Lambda}}\mathbf{U}^{\mathrm{T}}$, where $\tilde{\boldsymbol{\Lambda}}$ is $\boldsymbol{\Lambda}$ with all nonpositive eigenvalues replaced by positive entries (e.g. $|\Lambda_{ii}|$). Using the perturbed version in (5.5), results in a step of the form (5.4), so if we are not at a turning point, then a sufficiently small step in the direction $\boldsymbol{\Delta}$ is guaranteed to reduce f.
2. The second-order Taylor approximation about a point far from $\hat{\boldsymbol{\theta}}$ could be poor at $\hat{\boldsymbol{\theta}}$, so that there is no guarantee that stepping to its minimum will lead to a reduction in f. However, given the previous modification, we know that the Newton step is a descent direction. A small enough step in direction $\boldsymbol{\Delta}$ must reduce f. Therefore, if $f(\boldsymbol{\theta}' + \boldsymbol{\Delta}) > f(\boldsymbol{\theta}')$, repeatedly set $\boldsymbol{\Delta} \leftarrow \boldsymbol{\Delta}/2$, until a reduction in f is achieved.

[3] Positive definiteness can be tested by attempting a Choleski decomposition of the matrix concerned: it will succeed if the matrix is positive definite and fail otherwise. Use a pivoted Choleski decomposition to test for positive semi-definiteness. Alternatively, simply examine the eigenvalues returned by any symmetric eigen routine.

With these two modifications each step of Newton's method must reduce f until a turning point is reached.

In summary, starting with $k = 0$ and a guesstimate $\boldsymbol{\theta}^{[0]}$, iterate these steps:

1. Evaluate $f(\boldsymbol{\theta}^{[k]})$, $\nabla f(\boldsymbol{\theta}^{[k]})$ and $\nabla^2 f(\boldsymbol{\theta}^{[k]})$.
2. Test whether $\boldsymbol{\theta}^{[k]}$ is a minimum using (5.3), and terminate if it is.[4]
3. If $\mathbf{H} = \nabla^2 f(\boldsymbol{\theta}^{[k]})$ is not positive definite, perturb it so that it is.
4. Solve $\mathbf{H}\boldsymbol{\Delta} = -\nabla f(\boldsymbol{\theta}^{[k]})$ for the search direction $\boldsymbol{\Delta}$.
5. If $f(\boldsymbol{\theta}^{[k]} + \boldsymbol{\Delta})$ is not $< f(\boldsymbol{\theta}^{[k]})$, repeatedly halve $\boldsymbol{\Delta}$ until it is.
6. Set $\boldsymbol{\theta}^{[k+1]} = \boldsymbol{\theta}^{[k]} + \boldsymbol{\Delta}$, increment k by one and return to step 1.

In practice ∇f is not tested for exact equality to zero at 2, and we instead test whether $\|\nabla f(\boldsymbol{\theta}^{[k]})\| < |f(\boldsymbol{\theta}^{[k]})|\epsilon_r + \epsilon_a$, for small constants ϵ_r and ϵ_a.

Newton's method examples

As a single-parameter example, consider an experiment on antibiotic efficacy. A 1-litre culture of 5×10^5 cells is set up and dosed with antibiotic. After 2 hours, and then every subsequent hour up to 14 hours after dosing, 0.1ml of the culture is removed and the live bacteria in this sample counted under a microscope, giving counts, y_i, and times, t_i (hours). The data are

t_i	2	3	4	5	6	7	8	9	10	11	12	13	14
y_i	35	33	33	39	24	25	18	20	23	13	14	20	18

A simple model for the sample counts, y_i, is that their expected value is $E(Y_i) = \mu_i = 50e^{-\delta t_i}$, where δ is an unknown 'death rate' parameter (per hour) and t_i is the sample time in hours. Given the sampling protocol, it is reasonable to assume that the counts are observations of independent $\text{Poi}(\mu_i)$ random variables (see Section A.3.2), with probability function $f(y_i) = \mu_i^{y_i} e^{-\mu_i}/y_i!$ So the log likelihood is

$$
\begin{aligned}
l(\delta) &= \sum_{i=1}^{n} \{y_i \log(\mu_i) - \mu_i - \log(y_i!)\} \\
&= \sum_{i=1}^{n} y_i\{\log(50) - \delta t_i\} - \sum_{i=1}^{n} 50e^{-\delta t_i} - \sum_{i=1}^{n} \log(y_i!),
\end{aligned}
$$

[4] If the objective function contains a saddlepoint, then theoretically Newton's method might find it, in which case the gradient would be zero and the Hessian indefinite: in this rare case further progress can only be made by perturbing the Newton step directly.

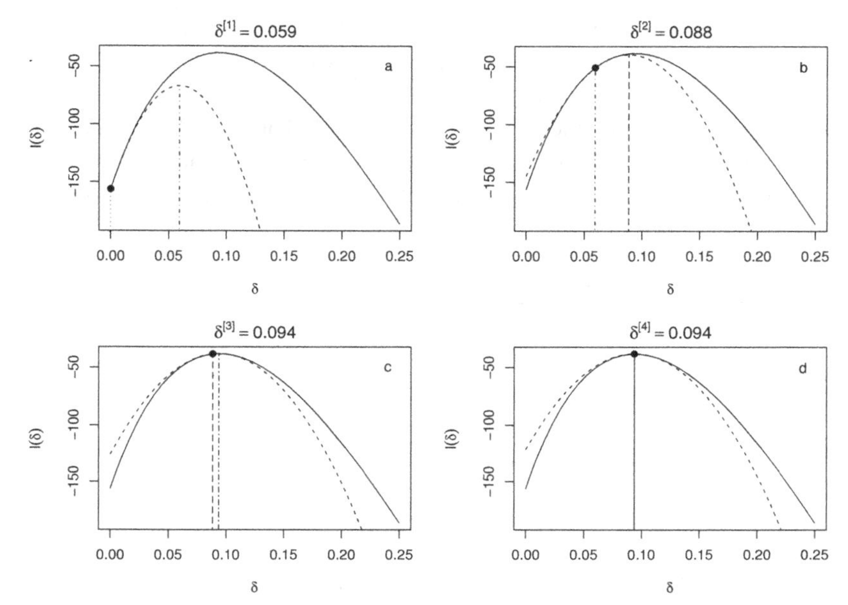

Figure 5.1 Newton's method for the antibiotic example of Section 5.1.1. Each panel shows one step of Newton's method, with the log likelihood (black) and the second-order Taylor approximation about • (dashed). Vertical lines show the estimated value of δ at the start and end of the step. Each panel title gives the end estimate. **a** starts from $\delta^{[0]} = 0$. **b** to **d** show subsequent iterations until convergence.

where $n = 13$. Differentiating w.r.t. δ,

$$\frac{\partial l}{\partial \delta} = -\sum_{i=1}^{n} y_i t_i + \sum_{i=1}^{n} 50 t_i e^{-\delta t_i} \text{ and } \frac{\partial^2 l}{\partial \delta^2} = -50 \sum_{i=1}^{n} t_i^2 e^{-\delta t_i}.$$

The presence of the t_i term in $e^{-\delta t_i}$ precludes a closed-form solution for $\partial l/\partial \delta = 0$, and Newton's method can be applied instead. Figure 5.1 illustrates the method's progression. In this case the second derivatives do not require perturbation, and no step-length halving is needed.

Now consider an example with a vector parameter. The following data are reported AIDS cases in Belgium, in the early stages of the epidemic.

Year (19–)	81	82	83	84	85	86	87	88	89	90	91	92	93
Cases	12	14	33	50	67	74	123	141	165	204	253	246	240

One important question, early in such epidemics, is whether control measures are beginning to have an impact or whether the disease is continuing to spread essentially unchecked. A simple model for unchecked growth leads to an 'exponential increase' model. The model says that the number of cases, y_i, is an observation of an independent Poisson r.v., with expected value $\mu_i = \alpha e^{\beta t_i}$ where t_i is the number of years since 1980. So the log likelihood is

$$l(\alpha, \beta) = \sum_{i=1}^{n} y_i\{\log(\alpha) + \beta t_i\} - \sum_{i=1}^{n}(\alpha e^{\beta t_i} - y_i!),$$

and hence,

$$\nabla l = \begin{bmatrix} \sum y_i/\alpha - \sum \exp(\beta t_i) \\ \sum y_i t_i - \alpha \sum t_i \exp(\beta t_i) \end{bmatrix}$$
$$\text{and } \nabla^2 l = \begin{bmatrix} -\sum y_i/\alpha^2 & -\sum t_i e^{\beta t_i} \\ -\sum t_i e^{\beta t_i} & -\alpha \sum t_i^2 e^{\beta t_i}. \end{bmatrix}.$$

A swift glance at the expression for the gradients should be enough to convince you that numerical methods will be required to find the MLEs of the parameters. Starting from an initial guess $\alpha^{[0]} = 4$, $\beta^{[0]} = .35$, here is the first Newton iteration:

$$\begin{bmatrix} \alpha^{[0]} \\ \beta^{[0]} \end{bmatrix} = \begin{bmatrix} 4 \\ .35 \end{bmatrix} \Rightarrow \nabla l = \begin{bmatrix} 88.4372 \\ 1850.02 \end{bmatrix},$$
$$\nabla^2 l = \begin{bmatrix} -101.375 & -3409.25 \\ -3409.25 & 154567 \end{bmatrix} \Rightarrow (\nabla^2 l)^{-1} \nabla l = \begin{bmatrix} -1.820 \\ 0.028 \end{bmatrix}$$
$$\Rightarrow \begin{bmatrix} \alpha^{[1]} \\ \beta^{[1]} \end{bmatrix} = \begin{bmatrix} \alpha^{[0]} \\ \beta^{[0]} \end{bmatrix} - (\nabla^2 l)^{-1} \nabla l = \begin{bmatrix} 5.82 \\ 0.322 \end{bmatrix}.$$

After eight more steps the likelihood is maximised at $\hat{\alpha} = 23.1$, $\hat{\beta} = 0.202$. Figure 5.2 illustrates six Newton steps, starting from the more interesting point $\hat{\alpha}_0 = 4$, $\hat{\beta}_0 = 0.35$. Perturbation to positive definiteness is required in the first two steps, but the method converges in six steps.

Newton variations: avoiding f evaluation and the expected Hessian

Occasionally we have access to ∇f and $\nabla^2 f$, but f itself is either unavailable or difficult to compute in a stable way. Newton's method only requires evaluation of f in order to check that the Newton step has led to a reduction in f. It is usually sufficient to replace the condition $f(\boldsymbol{\theta} + \boldsymbol{\Delta}) \le f(\boldsymbol{\theta})$ with the condition that f must be non-increasing in the direction $\boldsymbol{\Delta}$ at

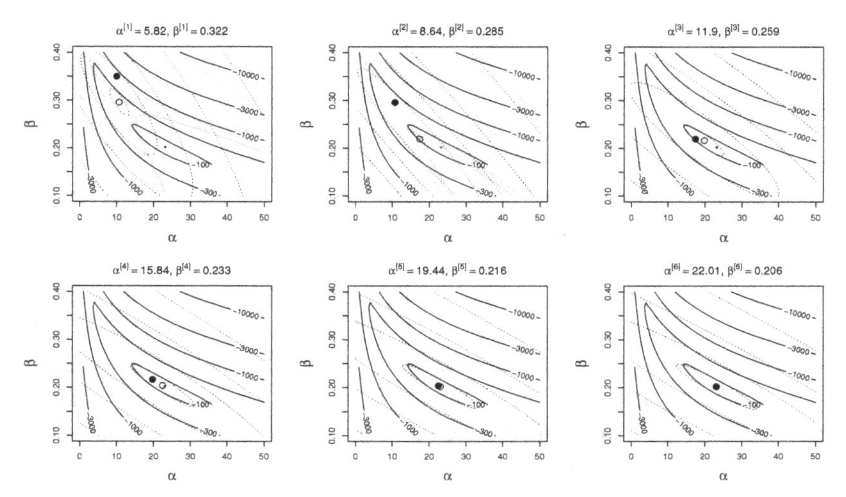

Figure 5.2 Newton's method for the AIDS example of Section 5.1.1. Each panel shows one step, with the log likelihood contoured in black, the second order Taylor approximation about • in grey, and the quadratic given by the positive definite corrected Hessian as dotted. ∘ gives the Newton method proposed parameter values at the end of each step, which are also given in the panel caption. The iteration starts at the top left and has converged by the lower right.

$\boldsymbol{\theta}' + \boldsymbol{\Delta}$. That is, $\nabla f(\boldsymbol{\theta}' + \boldsymbol{\Delta})^{\mathrm{T}}\boldsymbol{\Delta} \leq 0$. In many circumstances, step-length control based on this condition ensures convergence in cases where the iteration would otherwise have diverged, but unlike the function-value based control, pathological cases can easily be dreamt up to defeat it. Such step-length reduction should only be applied after testing that the step has not already met the convergence criteria. See Section 5.4.3 for an example.

Another common variation on the method, used in maximum likelihood estimation, is to replace $-\nabla^2 l(\boldsymbol{\theta})$ by $-E\{\nabla^2 l(\boldsymbol{\theta})\}$ (so-called *Fisher scoring*). Because the replacement is always positive (semi-)definite, perturbation to positive definiteness is not required and by the arguments surrounding (5.4) the method converges when used with simple step-length control.

5.1.2 Quasi-Newton

Newton's method is very effective and in the maximum likelihood setting has the nice property of being based on exactly the derivatives of the log likelihood that are required to use the large sample result (2.3). However, there are cases where the derivative vector ∇f is available, but the Hessian matrix, $\nabla^2 f$, is tedious or difficult to evaluate. If the dimension of $\boldsymbol{\theta}$ is large, numerically solving for the Newton direction can also become prohibitively costly. These considerations raise the question of what can be done based only on f and ∇f.

The obvious approach is to apply the strategy that led to Newton's method again, but based on the first-order Taylor expansion of f. However, the resulting *steepest descent* method is really a nonstarter. The problem is that the first-order Taylor expansion ceases to be a good model of f at exactly the point we are most interested in. At the minimum of f, $\nabla f = \mathbf{0}$ and we lose all justification for having neglected the second-order terms in the Taylor expansion in favour of the first order terms, since the latter have vanished. This theoretical concern is borne out in practice: the steepest descent method often becomes excruciatingly slow as it approaches a minimum.

A less obvious approach is to build up a local quadratic model of f from the first derivative information accumulated as the optimisation proceeds. This leads to *quasi-Newton* methods, which update an approximation to the Hessian, $\nabla^2 f$, based entirely on evaluations of ∇f. In principle, this approximation can be used instead of the Hessian in Newton's method, but it is also possible to work directly on an approximation to the inverse of the Hessian, thereby reducing the cost of calculating the step, $\boldsymbol{\Delta}$. It is also possible to ensure that the approximate Hessian is *always* positive definite.

Quasi-Newton methods were invented in the mid 1950s by W. C. Davidon (a physicist). In the mathematical equivalent of not signing the Beatles, his paper on the method was rejected (it was eventually published in 1991). There are now many varieties of the quasi-Newton method, but the most popular is the BFGS variant,[5] which is briefly covered here.

Suppose that $\mathbf{H}^{[k+1]}$ is the approximate positive definite Hessian at the $(k+1)^{\text{th}}$ step, so that

$$f(\boldsymbol{\theta}) \simeq f(\boldsymbol{\theta}^{[k+1]}) + \nabla f(\boldsymbol{\theta}^{[k+1]})^{\mathrm{T}}(\boldsymbol{\theta} - \boldsymbol{\theta}^{[k+1]}) + \frac{1}{2}(\boldsymbol{\theta} - \boldsymbol{\theta}^{[k+1]})^{\mathrm{T}}\mathbf{H}^{[k+1]}(\boldsymbol{\theta} - \boldsymbol{\theta}^{[k+1]}).$$

The basic requirement of a quasi Newton method is that this approximation should exactly match $\nabla f(\boldsymbol{\theta}^{[k]})$; that is, it should get the gradient vector at

[5] BFGS is named after Broyden, Fletcher, Goldfarb and Shanno all of whom discovered and published it, independently, around 1970. 'Big Friendly Giant Steps' is the way all Roald Dahl readers remember the name, of course (M. V. Bravington, pers. com.).

the previous point, $\boldsymbol{\theta}^{[k]}$, exactly right. So

$$\nabla f(\boldsymbol{\theta}^{[k+1]}) + \mathbf{H}^{[k+1]}(\boldsymbol{\theta}^{[k]} - \boldsymbol{\theta}^{[k+1]}) = \nabla f(\boldsymbol{\theta}^{[k]}),$$

which can be compactly re-written as

$$\mathbf{H}^{[k+1]}\mathbf{s}_k = \mathbf{y}_k, \tag{5.6}$$

where $\mathbf{s}_k = \boldsymbol{\theta}^{[k+1]} - \boldsymbol{\theta}^{[k]}$ and $\mathbf{y}_k = \nabla f(\boldsymbol{\theta}^{[k+1]}) - \nabla f(\boldsymbol{\theta}^{[k]})$. Equation (5.6) will only be feasible for positive definite $\mathbf{H}^{[k+1]}$ under certain conditions on $\mathbf{s}_k$ and $\mathbf{y}_k$, but these can always be met by choosing the step-length to meet the Wolfe conditions, covered shortly.

Now let us work in terms of the inverse approximate Hessian, $\mathbf{B}^{[k]} \equiv \left(\mathbf{H}^{[k]}\right)^{-1}$. Equation (5.6) alone does not define a unique $\mathbf{B}^{[k+1]}$, and some extra conditions are needed. A $\mathbf{B}^{[k+1]}$ is sought that

1. satisfies (5.6) so that $\mathbf{B}^{[k+1]}\mathbf{y}_k = \mathbf{s}_k$;
2. is as close as possible to $\mathbf{B}^{[k]}$;
3. is positive definite.

'Close' in condition 2 is judged using a particular matrix norm, that is not covered here. The unique solution to this problem is the BFGS update

$$\mathbf{B}^{[k+1]} = (\mathbf{I} - \rho_k \mathbf{s}_k \mathbf{y}_k^{\mathrm{T}})\mathbf{B}^{[k]}(\mathbf{I} - \rho_k \mathbf{y}_k \mathbf{s}_k^{\mathrm{T}}) + \rho_k \mathbf{s}_k \mathbf{s}_k^{\mathrm{T}},$$

where $\rho_k^{-1} = \mathbf{s}_k^{\mathrm{T}}\mathbf{y}_k$. The BFGS method then works exactly like Newton's method, but with $\mathbf{B}^{[k]}$ in place of the inverse of $\nabla^2 f(\boldsymbol{\theta}^{[k]})$, and without the need for evaluation of second derivatives or for perturbing the Hessian to achieve positive definiteness. A finite difference approximation to the Hessian is often used to start the method (see Section 5.5.2).

The only detail not required by Newton's method is that step-length selection must now be carried out more carefully. We must ensure that the step-length is such that $\boldsymbol{\Delta}$ satisfies the sufficient decrease condition

$$f(\boldsymbol{\theta}^{[k]} + \boldsymbol{\Delta}) \le f(\boldsymbol{\theta}^{[k]}) + c_1 \nabla f(\boldsymbol{\theta}^{[k]})^{\mathrm{T}}\boldsymbol{\Delta},$$

$c_1 \in (0, 1)$, and the curvature condition

$$\nabla f(\boldsymbol{\theta}^{[k]} + \boldsymbol{\Delta})^{\mathrm{T}}\boldsymbol{\Delta} \ge c_2 \nabla f(\boldsymbol{\theta}^{[k]})^{\mathrm{T}}\boldsymbol{\Delta},$$

$c_2 \in (c_1, 1)$. Collectively these conditions are known as the Wolfe conditions. The first seeks to ensure that the step results in a decrease that is reasonable relative to the gradient of f in the direction of $\boldsymbol{\Delta}$ and guards against overly long steps. The second says that there should have been a sufficient decrease in the gradient of the function along $\boldsymbol{\Delta}$ (otherwise why

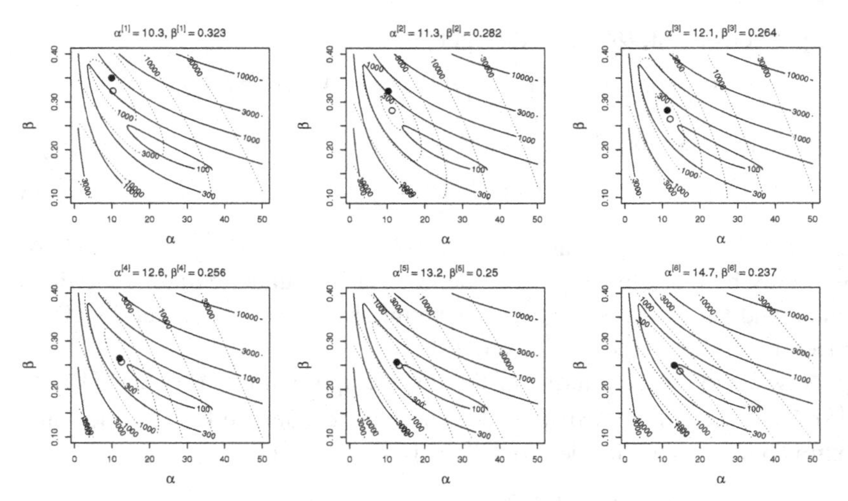

Figure 5.3 BFGS quasi-Newton method applied to the AIDS example of Section 5.1.1. Each panel shows one step, with the negative log likelihood as black contours, and the quadratic implied by the current gradient and approximate inverse Hessian about • as dashed contours. ∘ gives the updated parameter values at the end of each step, which are given numerically in each panel caption. The iteration starts at the top left. The MLE is reached in about six more steps.

not take a longer step, given that the function is still decreasing fast in this direction). For a full discussion see Nocedal and Wright (2006, §3.1), where $c_1 = 10^{-4}$ and $c_2 = 0.9$ are suggested as typical.

When performing maximum likelihood estimation, it is tempting to use the converged $\mathbf{B}$ matrix as an estimate of $\mathcal{I}^{-1}$, but caution is required. Because of the arguments surrounding (5.4), quasi Newton methods work even when $\mathbf{B}$ is a poor approximation to the inverse Hessian of f. $\mathbf{B}$ may be a poor representation of the shape of f in directions that the BFGS iteration has not explored recently.

Quasi-Newton example

Figure 5.3 illustrates the first 6 steps of BFGS applied to the AIDS in Belgium model of Section 5.1.1. Compared to the Newton method in Figure 5.2, progress is slightly slower, but convergence is still reached in about 12 steps, despite only requiring function values and first derivatives to be evaluated.

5.1.3 The Nelder-Mead polytope method

What if even gradient evaluation is too taxing, or if our objective is not smooth enough for Taylor approximations to be valid? What can be done with function values alone? The Nelder-Mead polytope[6] method provides an elegant answer.

Let p be the dimension of $\boldsymbol{\theta}$. At each stage of the method we maintain $p + 1$ distinct $\boldsymbol{\theta}$ vectors, defining a polytope in the parameter space (e.g. for a two-dimensional $\boldsymbol{\theta}$, the polytope is a triangle). The following steps are iterated until a minimum is reached/ the polytope collapses to a point.

1. The search direction is defined as the vector from the worst point (the vertex of the polytope with the highest objective value) through the average of the remaining p points.
2. The initial step-length is set to twice the distance from the worst point to the centroid of the others. If it succeeds (meaning that the new point is no longer the worst point), then a step-length of 1.5 times that is tried, and the better of the two accepted.
3. If the previous step did not find a successful new point, then step-lengths of half and one and a half times the distance from the worst point to the centroid are tried.
4. If the last two steps failed to locate a successful point, then the polytope is reduced in size by linear rescaling towards the current best point (which remains fixed.)

Variations are possible, in particular with regard to the step-lengths and shrinkage factors. Figure 5.4 illustrates the polytope method applied to the negative log likelihood of the AIDS data example of Section 5.1.1. Each polytope is plotted, with the line style cycling through, black, grey and dashed black. The worst point in each polytope is highlighted with a circle.

In this case it took 24 steps to reach the MLE. This is a somewhat higher number of steps than the Newton or BFGS methods, but given that we need no derivatives in this case, the amount of computation is actually less.

On the basis of this example you might be tempted to suppose that Nelder-Mead is all you ever need, but this is generally not the case. If you need to know the optimum very accurately (for example, for the inner optimisation in a nested optimisation), then Nelder-Mead will often take a long time to get an answer that Newton based methods would give very quickly. Also, the polytope can get 'stuck', so that it is usually a good idea to restart

[6] also known as the *downhill simplex method*, but not to be confused with the completely different *simplex method* of linear programming.

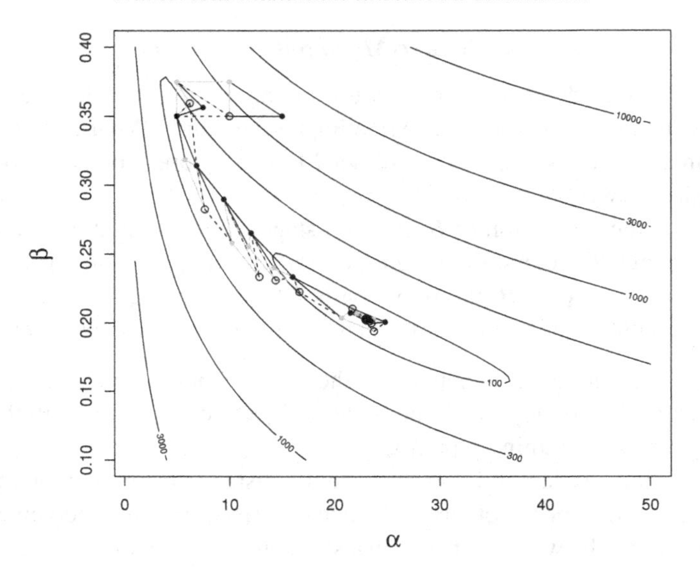

Figure 5.4 The Nelder-Mead method applied to the AIDS example of Section 5.1.1. All 24 steps to convergence are shown, from a starting point at $\alpha = 10$, $\beta = 0.35$. Polytope (triangle) line styles cycle through black, grey and dashed black. The worst vertex of each polytope is highlighted with a symbol.

the optimisation from any apparent minimum (with a new polytope having the apparent optimum as one vertex), to check that further progress is really not possible. The Nelder-Mead method is good if the answer does not need to be too accurate and derivatives are hard to come by.

5.2 A likelihood maximisation example in R

Echinus affinis is a species of deep sea urchin. Gage and Tyler (1985) reported data on the growth of *E. affinis* collected from the Rockall Trough, which are shown in Figure 5.5. Gurney and Nisbet (1998) suggested simple energy-budget based arguments to arrive at a model for volume, V, as a function of age, a, which is

$$\frac{\mathrm{d}V}{\mathrm{d}a} = \begin{cases} \gamma V & V < \phi/\gamma \\ \phi & \text{otherwise,} \end{cases}$$

where γ and ϕ are parameters. The initial volume is ω, also a model parameter. Growth is in two phases: in the first the animal grows as fast as it can, given the food it can obtain, and in the second it grows less quickly, putting

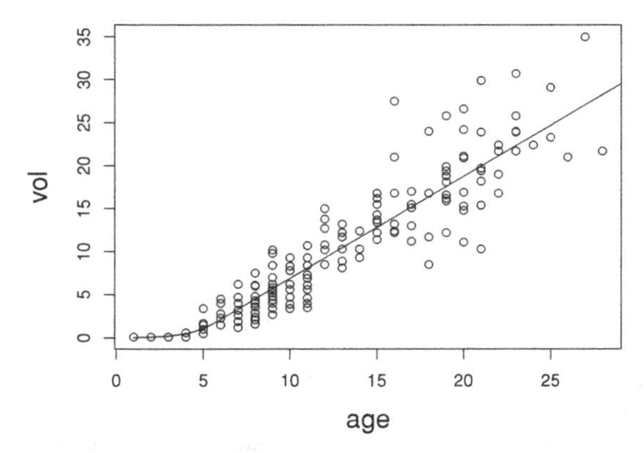

Figure 5.5 Data on urchin volume against age. Symbols show raw data. The black curve is the best fit model from Section 5.2.

the surplus food energy into reproduction. The age at onset of reproduction is therefore

$$a_m = \frac{1}{\gamma} \log \left(\frac{\phi}{\gamma\omega} \right),$$

and the model can be solved analytically:

$$V(a) = \begin{cases} \omega \exp(\gamma a) & a < a_m \\ \phi/\gamma + \phi(a - a_m) & \text{otherwise.} \end{cases}$$

Clearly the data do not follow the model exactly, so denoting the ith volume measurement as v_i, one possible model is that $\sqrt{v_i} \sim N(\sqrt{V(a_i)}, \sigma^2)$ where the observations are independent (a reasonable assumption because each datum relates to one individual).

5.2.1 Maximum likelihood estimation

Given the model specification, it is straightforward to code up an R function evaluating the negative log likelihood of $\boldsymbol{\theta} = \log(\omega, \gamma, \phi, \sigma)^{\mathrm{T}}$, where the log parameterisation ensures that ω, γ, ϕ and σ remain positive:

```
urchin.vol <- function(theta,age) {
## get volumes at 'age' given log params in 'theta'
  omega <- exp(theta[1]); gamma <- exp(theta[2])
  phi <- exp(theta[3]); V <- age*0
  am <- log(phi/(gamma*omega))/gamma
  ind <- age < am
  V[ind] <- omega*exp(gamma*age[ind])
```

```
  V[!ind] <- phi/gamma + phi*(age[!ind]-am)
  V
}

ll <- function(theta,age,vol) {
  rV <- sqrt(urchin.vol(theta,age)) ## expected sqrt vol.
  sigma <- exp(theta[4])
  -sum(dnorm(sqrt(vol),rV,sigma,log=TRUE)) ## -ve log lik.
}
```

Now let us minimise `ll` (i.e. maximise the log likelihood) with respect to $\boldsymbol{\theta}$, using the BFGS method available in R function `optim` (because no gradient function is supplied, `optim` will approximate gradients by finite differencing: see Section 5.5.2). Assume that the data are in a data frame, `uv`, with columns `vol` and `age`:

```
> th.ini <- c(0,0,0,0) ## initial parameter values
> fit <- optim(th.ini,ll,hessian=TRUE,method="BFGS",
+             age=uv$age,vol=uv$vol)
> fit
$par
[1] -4.0056322 -0.2128199 0.1715547 -0.7521029
$value
[1] 94.69095
$counts
function gradient
      74       25
$convergence
[1] 0
 ...
```

The first argument to `optim` provides initial parameter values from which to start the optimisation. The next argument is the objective function. The first argument of the objective function must be the vector of parameters with respect to which optimisation is required. The objective function may depend on other fixed arguments, which can be provided, named, via `optim`'s '`...`' argument: this is how `age` and `vol` get passed to `ll`. `hessian=TRUE` tells `optim` to return an approximate Hessian matrix at convergence, and `method="BFGS"` selects the BFGS optimisation method. The default method is Nelder-Mead.

The returned object, `fit`, contains several elements. `par` contains the minimising parameter values; here the MLE, $\hat{\boldsymbol{\theta}}$. `value` contains the value of the objective function at the minimum (the negative of the maximised log likelihood in this case). `counts` indicates how many function and gradient evaluations have been required (the latter by finite differencing, in this case). `convergence` contains a numeric code: `0` for convergence or other

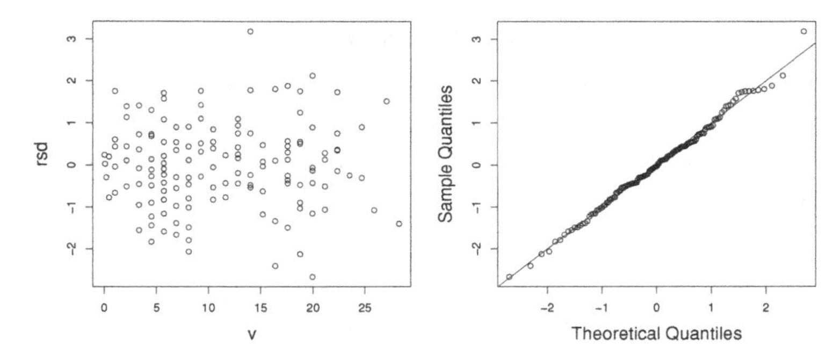

Figure 5.6 Checking plots for the urchin model from Section 5.2. Left: standardised residuals against fitted values. The lack of pattern in the mean or variance suggests that the assumptions are reasonable. Right: normal QQ-plot of standardised residuals; it is close enough to a straight line to accept the normality assumption.

integers indicating some problem (see `?optim`). `message` (not shown) contains any diagnostic message returned from the underlying optimisation code. `hessian` (not shown) contains the Hessian matrix.

5.2.2 Model checking

Before investigating the estimates further it is important to check that the model assumptions are plausible. Figure 5.5 overlays the estimated curve of $V(a)$ against a over the raw data. As a characterisation of the expected volume the model looks reasonable, but what of the distributional assumptions on which further inference is based? For this model is it easy to compute standardised residuals:

$$\hat{\epsilon}_i = \left\{ \sqrt{v_i} - \sqrt{V(a_i)} \right\} / \sigma,$$

which should be close to i.i.d. $N(0, 1)$ deviates if the model is correct. Then a plot of residuals against fitted values and a normal QQ-plot are useful.

```
theta <- fit$par ## MLE
v <- urchin.vol(theta,uv$age)
rsd <- (uv$vol^.5-v^.5)/exp(theta[4])
plot(v,rsd);qqnorm(rsd);abline(0,1);
```

The results are shown in Figure 5.6, and suggest no problem with the distributional assumptions.

5.2.3 Further inference

Now let us use (4.5) to obtain approximate 95% confidence intervals for the model parameters on the original scale. Notice that we already have the Hessian of the *negative* log likelihood, and at this level of approximation $1.96 \approx 2$.

```
> V <- solve(fit$hessian) ## approx. cov matrix
> sd <- diag(V)^.5  ## standard errors
> ## now get 95% CIs for all 4 parameters...
> rbind(exp(theta - 2*sd),exp(theta + 2*sd))
             [,1]      [,2]     [,3]      [,4]
[1,] 7.049868e-05 0.2138756 1.092046 0.4186246
[2,] 4.705125e+00 3.0548214 1.290534 0.5307708
```

The intervals for ω and γ are very wide, but computing the estimated correlation matrix of $\hat{\boldsymbol{\theta}}$ using `diag(1/sd)%*%V%*%(diag(1/sd))`, we find a correlation of 0.997 between $\log \omega$ and $\log \gamma$, which explains the width.

As a simple example of model selection, suppose that we want to test the hypothesis that $\sigma = 0.4$. The above interval for σ suggests rejecting this hypothesis at the 5% level, but a generalised likelihood ratio test could also be used for this purpose, using (2.4). To implement this we need a modified version of `ll`, `ll0`, say, in which the line `sigma <-exp(theta[4])` is removed and replaced with a function argument `sigma = 0.4`, so that the null hypothesis is imposed. It remains to optimise `ll0`, evaluate the log likelihood ratio statistic and compute the p-value using (2.4):

```
> fit0 <- optim(rep(0,3),ll0,method="BFGS",
+                 age=uv$age,vol=uv$vol,sigma=0.4)
> llr <- fit0$value - fit$value
> pchisq(2*llr,1,lower.tail=FALSE)
[1] 0.003421646
```

This suggests rather more evidence against the null than the interval might have implied. We could also search for the range of σ values acceptable in a generalised likelihood ratio test. The following code does this.

```
llf <- function(sigma,ll.max,uv) # zero on accept boundary
    -2*(ll.max-optim(rep(0,3),ll0,method="BFGS",age=uv$age,
       vol=uv$vol,sigma=sigma)$value)-qchisq(.95,1)
uniroot(llf,c(.2,.47),uv=uv,ll.max=fit$value)$root # lower
uniroot(llf,c(.47,1),uv=uv,ll.max=fit$value)$root # upper
```

The resulting 95% *profile likelihood* interval for σ is (0.421, 0.532).

AIC also suggests that the simplified model is not as good as the original:

```
> 2*fit$value + 2*4; 2*fit0$value + 2*3 ## AIC
[1] 197.3819
[1] 203.9497
```

However, in its current form the model is somewhat unbiological. Measurement errors in the urchin volumes are likely to be rather small, and much of the observed variability is likely to result from variation between individuals in their realized growth parameters γ and ϕ, suggesting a reformulation with random effects.

5.3 Maximum likelihood estimation with random effects

When random effects are present it is usually straightforward to write down the joint density, $f_\theta(\mathbf{y}, \mathbf{b})$, of the observed data, $\mathbf{y}$, and unobserved random effects, $\mathbf{b}$, which depends on parameters $\boldsymbol{\theta}$. However, the likelihood is the marginal density of the data evaluated at the observed data values,

$$L(\boldsymbol{\theta}) = f_\theta(\mathbf{y}) = \int f_\theta(\mathbf{y}, \mathbf{b}) d\mathbf{b}, \tag{5.7}$$

and the integral is usually analytically intractable. We then have several options:

1. Use numerical integration (also known as 'quadrature'). This is usually impractical unless the integral can be decomposed into a product of low-dimensional integrals or $\mathbf{b}$ is low-dimensional.
2. Estimate the integral by Monte Carlo methods. This can be effective, but is not always easy to combine with numerical likelihood maximisation, and accuracy considerations mean that we must typically simulate many times as many $\mathbf{b}$ values as we have data in $\mathbf{y}$.
3. Approximate the integral with one that we can do.
4. Avoid the integral altogether by finding an easier to evaluate function whose maximum will coincide with the maximum of the likelihood.

The following sections consider options 3 and 4, looking in particular at Laplace approximation, the EM algorithm, and the combination of the two.

5.3.1 Laplace approximation

Let $\hat{\mathbf{b}}_y$ be the value of $\mathbf{b}$ maximising $f(\mathbf{y}, \mathbf{b})$ for a given $\mathbf{y}$ (the dependence on $\boldsymbol{\theta}$ has been dropped from the notation to avoid clutter). Then a second-order Taylor expansion of $\log f$, about $\hat{\mathbf{b}}_y$, gives

$$\log f(\mathbf{y}, \mathbf{b}) \simeq \log f(\mathbf{y}, \hat{\mathbf{b}}_y) - \frac{1}{2}(\mathbf{b} - \hat{\mathbf{b}}_y)^{\mathrm{T}}\mathbf{H}(\mathbf{b} - \hat{\mathbf{b}}_y),$$

where $\mathbf{H} = -\nabla_b^2 \log f(\mathbf{y}, \hat{\mathbf{b}}_y)$. Hence,

$$f(\mathbf{y}, \mathbf{b}) \simeq f(\mathbf{y}, \hat{\mathbf{b}}_y) \exp\left\{-\frac{1}{2}(\mathbf{b} - \hat{\mathbf{b}}_y)^{\mathrm{T}}\mathbf{H}(\mathbf{b} - \hat{\mathbf{b}}_y)\right\}.$$

However, writing n_b for the dimension of $\mathbf{b}$,

$$\int \frac{1}{(2\pi)^{n_b/2}|\mathbf{H}^{-1}|^{1/2}} \exp\left\{-\frac{1}{2}(\mathbf{b} - \hat{\mathbf{b}}_y)^{\mathrm{T}}\mathbf{H}(\mathbf{b} - \hat{\mathbf{b}}_y)\right\} d\mathbf{b} = 1,$$

since the integrand is the p.d.f. of an $N(\hat{\mathbf{b}}_y, \mathbf{H}^{-1})$ random vector and p.d.f.s integrate to 1. It follows that

$$\int f(\mathbf{y}, \mathbf{b}) d\mathbf{b} \simeq f(\mathbf{y}, \hat{\mathbf{b}}_y) \int \exp\left\{-\frac{1}{2}(\mathbf{b} - \hat{\mathbf{b}}_y)^{\mathrm{T}}\mathbf{H}(\mathbf{b} - \hat{\mathbf{b}}_y)\right\} d\mathbf{b}$$

$$= f(\mathbf{y}, \hat{\mathbf{b}}_y) \frac{(2\pi)^{n_b/2}}{|\mathbf{H}|^{1/2}}, \quad (5.8)$$

the right-hand side being the first order *Laplace approximation* to the integral. Careful accounting of the approximation error shows it to generally be $O(n^{-1})$ where n is the sample size (assuming a fixed length for $\mathbf{b}$).

Notice how the problem of evaluating the integral has been reduced to the problem of finding $\nabla^2 \log f(\mathbf{y}, \hat{\mathbf{b}}_y)$ and $\hat{\mathbf{b}}_y$. If we can obtain the former then the latter is always obtainable by Newton's method. Of course optimising the approximate likelihood that results from the Laplace approximation will also require numerical optimisation, so nested optimisation loops will usually be needed; but this is usually preferable to a brute force attack on (5.7).

5.3.2 The EM algorithm

A rather ingenious method avoids the integral in (5.7) altogether, replacing it with an integral that is sometimes more analytically tractable; in any case it can readily be approximated to greater accuracy than is straightforward for (5.7) itself. The method starts from a parameter guess, $\boldsymbol{\theta}'$, and the standard decomposition:

$$\log f_\theta(\mathbf{y}, \mathbf{b}) = \log f_\theta(\mathbf{b}|\mathbf{y}) + \log f_\theta(\mathbf{y}).$$

The idea is then to take the expectation of $\log f_\theta(\mathbf{y}, \mathbf{b})$ with respect to $f_{\theta'}(\mathbf{b}|\mathbf{y})$ (play close attention to when $\boldsymbol{\theta}$ is primed and when it is not). For some models this expectation can readily be computed, but otherwise we can approximate it to relatively high accuracy. In any case we obtain

$$E_{b|y,\theta'} \log f_\theta(\mathbf{y}, \mathbf{b}) = E_{b|y,\theta'} \log f_\theta(\mathbf{b}|\mathbf{y}) + \log f_\theta(\mathbf{y}),$$

which it is convenient to rewrite as

$$Q_{\theta'}(\boldsymbol{\theta}) = P_{\theta'}(\boldsymbol{\theta}) + l(\boldsymbol{\theta}) \tag{5.9}$$

by definition of Q and P and recognising that the final term is simply the log likelihood $l(\boldsymbol{\theta})$. Now precisely the same argument that leads to (4.4) in Section 4.1 implies that $E_{b|y,\theta'} \log f_\theta(\mathbf{b}|\mathbf{y})$ is maximised when $\boldsymbol{\theta} = \boldsymbol{\theta}'$. So $P_{\theta'}$ is maximised at $P_{\theta'}(\boldsymbol{\theta}')$. It follow that if $Q_{\theta'}(\boldsymbol{\theta}) > Q_{\theta'}(\boldsymbol{\theta}')$, then $l(\boldsymbol{\theta}) > l(\boldsymbol{\theta}')$, since we know that $P_{\theta'}(\boldsymbol{\theta}) < P_{\theta'}(\boldsymbol{\theta}')$. That is, any $\boldsymbol{\theta}$ value that increases $Q_{\theta'}(\boldsymbol{\theta})$ relative to $Q_{\theta'}(\boldsymbol{\theta}')$ must result from $l(\boldsymbol{\theta})$ having increased, because $P_{\theta'}(\boldsymbol{\theta})$ will have decreased. So any change we make to $\boldsymbol{\theta}$ that increases $Q_{\theta'}(\boldsymbol{\theta})$ must increase $l(\boldsymbol{\theta})$.

$Q_{\hat{\theta}}(\boldsymbol{\theta})$ has a maximum at $\hat{\boldsymbol{\theta}}$, because both $P_{\hat{\theta}}(\boldsymbol{\theta})$ and $l(\boldsymbol{\theta})$ have maxima at $\hat{\boldsymbol{\theta}}$. Further, $Q_{\theta'}(\boldsymbol{\theta})$ can only be maximised at $\boldsymbol{\theta}'$, if $l(\boldsymbol{\theta}')$ is a turning point: otherwise $P_{\theta'}(\boldsymbol{\theta})$ is maximised at $\boldsymbol{\theta}'$ while $l(\boldsymbol{\theta})$ is not.

Taken together these properties of Q imply that if we repeatedly find $\boldsymbol{\theta}^* = \text{argmax}_\theta\, Q_{\theta'}(\boldsymbol{\theta})$, and then set $\boldsymbol{\theta}' \leftarrow \boldsymbol{\theta}^*$, the resulting sequence of $\boldsymbol{\theta}'$ values leads to a monotonic increase in the likelihood and eventually converges on a turning point of $l(\boldsymbol{\theta})$: hopefully $\hat{\boldsymbol{\theta}}$. This iteration is known as the EM algorithm from the two steps of first obtaining the function $Q_{\theta'}(\boldsymbol{\theta})$ by taking an Expectation and then Maximising it with respect to $\boldsymbol{\theta}$.

In its basic form the EM algorithm is somewhat slow to converge when close to $\hat{\boldsymbol{\theta}}$, but Q also allows us to compute the gradient and Hessian of l from it, thereby facilitating the application of Newton's method to l, without actually evaluating l. Differentiating (5.9) with respect to $\boldsymbol{\theta}$ and evaluating at $\boldsymbol{\theta} = \boldsymbol{\theta}'$ we find that

$$\left.\frac{\partial Q_{\theta'}(\boldsymbol{\theta})}{\partial \boldsymbol{\theta}}\right|_{\theta=\theta'} = \left.\frac{\partial P_{\theta'}(\boldsymbol{\theta})}{\partial \boldsymbol{\theta}}\right|_{\theta=\theta'} + \left.\frac{\partial l(\boldsymbol{\theta})}{\partial \boldsymbol{\theta}}\right|_{\theta=\theta'} = \left.\frac{\partial l(\boldsymbol{\theta})}{\partial \boldsymbol{\theta}}\right|_{\theta=\theta'} \tag{5.10}$$

since $P_{\theta'}(\boldsymbol{\theta})$ has a maximum at $\boldsymbol{\theta} = \boldsymbol{\theta}'$, and hence its derivatives vanish. Some more work (e.g Davison, 2003, §5.5.2) establishes a result that is also useful with (2.3):

$$\left.\frac{\partial^2 l(\boldsymbol{\theta})}{\partial \boldsymbol{\theta} \partial \boldsymbol{\theta}^{\mathrm{T}}}\right|_{\theta=\theta'} = \left.\frac{\partial^2 Q_{\theta'}(\boldsymbol{\theta})}{\partial \boldsymbol{\theta} \partial \boldsymbol{\theta}^{\mathrm{T}}}\right|_{\theta=\theta'} + \left.\frac{\partial^2 Q_{\theta'}(\boldsymbol{\theta})}{\partial \boldsymbol{\theta} \partial \boldsymbol{\theta}'^{\mathrm{T}}}\right|_{\theta=\theta'}. \tag{5.11}$$

Equation (5.11) also enables maxima of $l(\boldsymbol{\theta})$ to be distinguished from other turning points that the EM algorithm might discover, since the Hessian will be indefinite in the latter cases (having positive and negative eigenvalues).

Higher order Laplace approximation for the E step

At this point the reader may reasonably object that the EM method requirement to evaluate

$$E_{b|y,\theta'} \log f_\theta(\mathbf{y}, \mathbf{b}) = \int \log f_\theta(\mathbf{y}, \mathbf{b}) f_{\theta'}(\mathbf{b}|\mathbf{y}) d\mathbf{b} \tag{5.12}$$

involves an integral that in general looks no more tractable than the integral (5.7) we are trying to avoid. In fact, there are many special cases where the expectation is much easier than (5.7), which is where the real strength of the approach lies. However it also turns out that a simple Laplace approximation (e.g. Steele, 1996) to (5.12) can be much more accurate than (5.8).

The key work is by Tierney et al. (1989), who consider approximation of conditional expectations of the form

$$E\{g(\mathbf{b})\} = \frac{\int g(\mathbf{b}) f_\theta(\mathbf{y}, \mathbf{b}) d\mathbf{b}}{\int f_\theta(\mathbf{y}, \mathbf{b}) d\mathbf{b}},$$

via-first order Laplace approximation of both integrals. This fails if g is not strictly positive, so consider estimating the moment generating function $M(s) = E[\exp\{sg(\mathbf{b})\}]$ and using $E(g) = \mathrm{d} \log M(s)/\mathrm{d}s|_{s=0}$.

$$\begin{aligned} M(s) &= \frac{\int \exp\{sg(\mathbf{b})\} f_\theta(\mathbf{y}, \mathbf{b}) d\mathbf{b}}{\int f_\theta(\mathbf{y}, \mathbf{b}) d\mathbf{b}} \\ &= \frac{\int \exp\{sg(\mathbf{b}) + \log f_\theta(\mathbf{y}, \mathbf{b})\} d\mathbf{b}}{\int \exp\{\log f_\theta(\mathbf{y}, \mathbf{b})\} d\mathbf{b}} = \frac{\int e^{h_s(\mathbf{b})} d\mathbf{b}}{\int e^{h(\mathbf{b})} d\mathbf{b}} \end{aligned}$$

by definition of h_s and h. Let $\hat{\mathbf{b}}$ maximise h and $\hat{\mathbf{b}}_s$ maximise h_s. Furthermore define $\mathbf{H} = -\nabla^2 h(\hat{\mathbf{b}})$ and $\mathbf{H}_s = -\nabla^2 h_s(\hat{\mathbf{b}}_s)$. A standard first order Laplace approximation of both integrals yields

$$\hat{M}(s) = \frac{|\mathbf{H}|^{1/2} f_\theta(\mathbf{y}, \hat{\mathbf{b}}_s) e^{sg(\hat{\mathbf{b}}_s)}}{|\mathbf{H}_s|^{1/2} f_\theta(\mathbf{y}, \hat{\mathbf{b}})}. \tag{5.13}$$

Tierney et al. (1989) show that the error in this approximation is $O(n^{-2})$ provided that h/n and h_s/n are of constant order (i.e. have magnitudes that do not depend on n). $\hat{E}(g) = \mathrm{d} \log \hat{M}(s)/\mathrm{d}s|_{s=0}$ is now the estimate of $E(g)$, which Tierney et al. (1989) also show has $O(n^{-2})$ error. Using the fact that $\hat{\mathbf{b}}_s = \hat{\mathbf{b}}$ when $s = 0$, and the fact that the first derivatives of

f_θ w.r.t. $\mathbf{b}_s$ are therefore zero at $s = 0$, then evaluating the derivative of the log of (5.13) at $s = 0$ gives

$$\hat{E}(g) = g(\hat{\mathbf{b}}) - \frac{1}{2}\frac{\mathrm{d}}{\mathrm{d}s}\log|\mathbf{H}_s|\bigg|_{s=0}. \tag{5.14}$$

To avoid requiring third derivatives w.r.t. $\mathbf{b}$, Tierney et al. (1989) suggest using centred finite differencing to approximate the derivative w.r.t. s. In the context of the EM algorithm, $g(\mathbf{b}) = \log f_\theta(\mathbf{y}, \mathbf{b})$, and all other quantities are evaluated at $\boldsymbol{\theta} = \boldsymbol{\theta}'$. So $h_s = s \log f_\theta(\mathbf{y}, \mathbf{b}) + \log f_{\theta'}(\mathbf{y}, \mathbf{b})$, and the approximation then gives us $Q_{\boldsymbol{\theta}'}(\boldsymbol{\theta})$ with $O(n^{-2})$ error.

5.4 R random effects MLE example

Consider again the urchin growth model from Section 5.2. A more biologically realistic model for the variability in these data might be that

$$V_i = \begin{cases} \omega \exp(g_i a_i) & a_i < a_{mi} \\ p_i/g_i + p_i(a_i - a_{mi}) & \text{otherwise} \end{cases}$$

where $a_{mi} = \log\{p_i/(g_i\omega)\}/g_i$, $\log g_i \sim N(\mu_g, \sigma_g^2)$ and $\log p_i \sim N(\mu_p, \sigma_p^2)$ (all independent), so that $\sqrt{v_i} \sim N(\sqrt{V_i}, \sigma^2)$. So in this model each urchin has its own growth rates, drawn from log-normal distributions, and the model parameters are ω, μ_g, σ_g, μ_p, σ_p and σ. Clearly the joint density of the data and random effects is easy to evaluate here, but the integral required to obtain the likelihood is intractable.

5.4.1 Direct Laplace approximation

To use Laplace approximation for the likelihood requires that we find the maximum of the log joint density of random effects and data w.r.t. the random effects, along with the corresponding Hessian. This entails writing a routine to evaluate the joint density, and its gradient and Hessian w.r.t. the random effects. An easy way to do this is to write out the joint density as an R expression and then have the `deriv` function do the heavy lifting (see Section 5.5.3). The only snag in the current case is that the urchins for which $a_i < a_{mi}$ have to be dealt with separately from the others.

Here is R code for producing a function `v0` with the arguments listed in `function.arg`, which will return predicted volumes for not yet mature urchins, given values for the parameters and random effects, as well as gradient and Hessian w.r.t. the random effects. Notice the use of log parameterisations here.

```
v0e <- expression(-log(2*pi*sigma^2)/2 -
     (sqrt(y) - sqrt(exp(w)*exp(exp(g)*a)))^2/(2*sigma^2)
     - log(2*pi) - log(sig.g*sig.p) -
     (g-mu.g)^2/(2*sig.g^2) - (p-mu.p)^2/(2*sig.p^2))

v0 <- deriv(v0e,c("g","p"), hessian=TRUE,function.arg=
     c("a","y","g","p","w","mu.g","sig.g","mu.p",
                                        "sig.p","sigma"))
```

Similarly tedious code produces a function `v1` for volumes of mature urchins. Only the expression for the mean volume changes to produce this. Next we need a function to evaluate the log joint density and its derivatives, suitable for optimising. Let `b` denote the vector containing the random effects: g_is first, then p_is. `y` is volume data and `a` contains the ages.

```
lfyb <- function(b,y,a,th) {
## evaluate joint p.d.f. of y and b + grad. and Hessian.
  n <- length(y)
  g <- b[1:n]; p <- b[1:n+n]
  am <- (p-g-th[1])/exp(g)
  ind <- a < am
  f0 <- v0(a[ind],y[ind],g[ind],p[ind],
          th[1],th[2],th[3],th[4],th[5],th[6])
  f1 <- v1(a[!ind],y[!ind],g[!ind],p[!ind],
          th[1],th[2],th[3],th[4],th[5],th[6])
  lf <- sum(f0) + sum(f1)
  g <- matrix(0,n,2) ## extract gradient to g...
  g[ind,] <- attr(f0,"gradient") ## dlfyb/db
  g[!ind,] <- attr(f1,"gradient") ## dlfyb/db
  h <- array(0,c(n,2,2)) ## extract Hessian to H...
  h[ind,,] <- attr(f0,"hessian")
  h[!ind,,] <- attr(f1,"hessian")
  H <- matrix(0,2*n,2*n)
  for (i in 1:2) for (j in 1:2) {
    indi <- 1:n + (i-1)*n; indj <- 1:n + (j-1)*n
    diag(H[indi,indj]) <- h[,i,j]
  }
  list(lf=lf,g=as.numeric(g),H=H)
}
```

The code for creating the full Hessian matrix `H` makes it clear that the Hessian is very sparse (mostly zeroes). What follows would be more efficient if the sparsity was exploited, but this would be a distraction at present.

The next step is to write an approximate log-likelihood function. Its main element is a loop to maximise the joint density w.r.t. the random effects, using Newton's method. Recall that to guarantee convergence we need to check the Hessian for positive definiteness at each step and perturb it if necessary. One way to do this is to check whether a Choleski decomposition of the Hessian is possible and to add a multiple of the identity matrix to it if needed. The Choleski factor also provides an efficient way of solving for the search direction, so the following function returns the Choleski factor of the Hessian or its positive definite modification:

```
pdR <- function(H,k.mult=20,tol=.Machine$double.eps^.8) {
  k <- 1; tol <- tol * norm(H); n <- ncol(H)
  while (inherits(try(R <- chol(H + (k-1)*tol*diag(n)),
            silent=TRUE),"try-error")) k <- k * k.mult
  R
}
```

Finally, here is the approximate negative log likelihood:

```
llu<-function(theta,vol,age,tol=.Machine$double.eps^.8){
## Laplace approximate log likelihood for urchin model.
  ii <- c(3,5,6)
  theta[ii] <- exp(theta[ii]) ## variance params
  n <- length(vol)
  if (exists(".inib",envir=environment(llu))) {
    b <- get(".inib",envir=environment(llu))
  } else b <- c(rep(theta[2],n),rep(theta[4],n)); ## init
  lf <- lfyb(b,vol,age,theta)
  for (i in 1:200) { ## Newton loop...
    R <- pdR(-lf$H) ## R'R = (perturbed) Hessian
    step <- backsolve(R,forwardsolve(t(R),lf$g)) ## Newton
    conv <- ok <- FALSE
    while (!ok) { ## step halving
      lf1 <- lfyb(b+step,vol,age,theta);
      if (sum(abs(lf1$g)>abs(lf1$lf)*tol)==0) conv <- TRUE
      kk <- 0
      if (!conv&&kk<30&&
          (!is.finite(lf1$lf) || lf1$lf < lf$lf)) {
        step <- step/2;kk <- kk+1
      } else ok <- TRUE
    }
    lf <- lf1;b <- b + step
    if (kk==30||conv) break ## if converged or failed
  } ## end of Newton loop
  assign(".inib",b,envir=environment(llu))
  R <- pdR(-lf$H,10)
  ll <- lf$lf - sum(log(diag(R))) + log(2*pi)*n
  -ll
}
```

We can save computer time by having `llu` save the maximising random effects, $\hat{\mathbf{b}}$, between calls, and use the previously stored $\hat{\mathbf{b}}$ as starting values next time it is called: this is achieved by the calls to `get` and `assign`, which store and retrieve $\hat{\mathbf{b}}$ from the environment of `llu`. Notice the assumption that a log parameterisation is used for the variance parameters.

Fitting can now be accomplished by `optim`, exactly as for the simpler likelihood in Section 5.2.

```
> th <- c(-4,-.2,log(.1),.2,log(.1),log(.5)) ## initial
> fit <- optim(th,llu,method="BFGS",vol=uv$vol,
+                  age=uv$age,hessian=TRUE)
> 2*fit$value + 2*length(fit$par) ## AIC
[1] 196.5785
```

So all the extra work for a more biologically plausible model has at least not increased the AIC.

5.4.2 EM optimisation

Now consider fitting the same model using the EM algorithm. Analytic evaluation of the expectation step does not look feasible, so let us use the method of Section 5.3.2. Direct Laplace approximation is based on differentiation and maximisation of the log joint density of random effects and data, $\log f_\theta(\mathbf{y}, \mathbf{b})$. Higher order approximation of the E-step requires the equivalent for $s \log f_\theta(\mathbf{y}, \mathbf{b}) + \log f_{\theta'}(\mathbf{y}, \mathbf{b})$, with arbitrary values of the constant s. Here is a function to evaluate this, with its gradient and Hessian.

```
lfybs <- function(s,b,vol,age,th,thp) {
## evaluate s log f(y,b;th) + log f(y,b;thp)
  lf <- lfyb(b,vol,age,thp)
  if (s!=0) {
    lfs <- lfyb(b,vol,age,th)
    lf$lf <- lf$lf + s * lfs$lf;lf$g <- lf$g + s * lfs$g
    lf$H <- lf$H + s * lfs$H
  }
  lf
}
```

Next we need a function to maximise this w.r.t. `b`. The following is really just a modification of `llu`, which returns $\log f_\theta(\mathbf{y}, \hat{\mathbf{b}})$ if $s = 0$, and $\log |\mathbf{H}_s|/2$ otherwise, in accordance with the ingredients needed to compute $Q_{\theta'}(\boldsymbol{\theta})$ using (5.14):

```
laplace <- function(s=0,th,thp,vol,age,b=NULL,
                    tol=.Machine$double.eps^.7) {
  ii <- c(3,5,6);thp[ii] <- exp(thp[ii])
  th[ii] <- exp(th[ii]) ## variance params
  n <- length(vol)
  ## initialize b ...
  if (is.null(b)) b <- c(rep(thp[2],n),rep(thp[4],n));
  lf <- lfybs(s,b,vol,age,th,thp)
  for (i in 1:200) { ## Newton loop to find b hat
```

```
    R <- pdR(-lf$H) ## R'R = fixed Hessian, R upper tri.
    step <- backsolve(R,forwardsolve(t(R),lf$g)) ## Newton
    conv <- ok <- FALSE
    while (!ok) {
      lf1 <- lfybs(s,b+step,vol,age,th,thp);
      if (sum(abs(lf1$g)>abs(lf1$lf)*tol)==0 ||
                     sum(b+step!=b)==0) conv <- TRUE
      kk <- 0
      if (!conv&&kk<30&&(!is.finite(lf1$lf) ||
                               lf1$lf < lf$lf)) {
        step <- step/2;kk <- kk+1
      } else ok <- TRUE
    }
    dlf <- abs(lf$lf-lf1$lf);lf <- lf1;b <- b + step;
    if (dlf<tol*abs(lf$lf)||conv||kk==30) break
  } ## end Newton loop
  if (s==0) {
    return(list(g=lfyb(b,vol,age,th)$lf,b=b))
  }
  R <- pdR(-lf$H,10)
  list(b=b,rldetH = sum(log(diag(R))))
}
```

The rest is straightforward. Here is a function to evaluate th.e Q function (again storing $\hat{\mathbf{b}}$ to use as starting values at the next call). The derivative required by (5.14) is obtained by finite differencing (see Section 5.5.2).

```
Q <- function(th,thp,vol,age,eps=1e-5) {
## 1. find b.hat maximising log joint density at thp
  if (exists(".inib",envir=environment(Q))) {
    b <- get(".inib",envir=environment(Q))
  } else b <- NULL
  la <- laplace(s=0,th,thp,vol,age,b=b)
  assign(".inib",la$b,envir=environment(Q))
## 2. For s = -eps and eps find b maximising s log joint
##   at th + log joint at thp along with log|H_s|.
  lap <- laplace(s=eps/2,th,thp,vol,age,b=la$b)$rldetH
  lam <- laplace(s= -eps/2,th,thp,vol,age,b=la$b)$rldetH
  la$g - (lap-lam)/eps
}
```

The basic EM iteration is now routine:

```
> thp <- th <- rep(0,6); ## starting values
> for (i in 1:30) { ## EM loop
+   er <- optim(th,Q,control=list(fnscale=-1,maxit=200),
+             vol=uv$vol,age=uv$age,thp=thp)
+   th <- thp <- er$par
+   cat(th,"\n")
+ }
-1.30807 -0.104484 0.015933 -0.351366 -0.422658 -0.22497
```

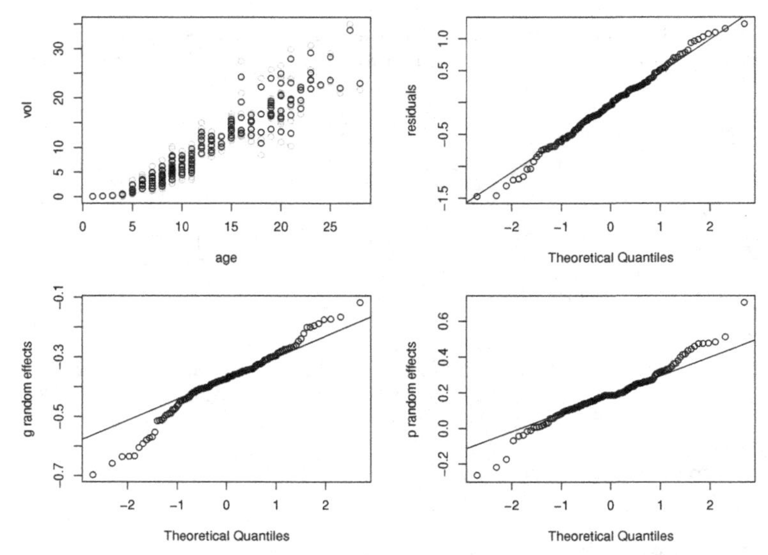

Figure 5.7 Checking plots for the full urchin model from Section 5.4. Top left: the predicted urchin volumes given the predicted random effects are shown in black, with the raw data in grey. Top right: Normal QQ-plot for the residual errors. Bottom left: Normal QQ-plot for predicted random effects $\hat{\mathbf{g}}$. Bottom right: Normal QQ-plot for predicted random effects $\hat{\mathbf{p}}$. Both random effects appear somewhat heavy tailed.

```
-1.13297 -0.220579 0.049261 -0.240472 -0.724219 -0.42390
                  [7 iterations omitted]
-2.91226 -0.162600 -1.079699 -0.049739 -1.247416 -1.27902
                  [19 iterations omitted]
-3.39816 -0.322957 -1.550822 0.150278 -1.512047 -1.37022
```

The Nelder-Mead method has been used to optimise Q, with a step limit of 200 to avoid excessive refinement of an optimum that will anyway be discarded at the next step. For the first few steps, far from the optimum, the algorithm makes good progress, but thereafter progress is slow, which is the main practical problem with the basic EM iteration. After the first few steps it is better to switch to Newton based optimisation by making use of (5.10) and (5.11).

5.4.3 EM-based Newton optimisation

Here is a simple routine to find derivatives of the log likelihood according to (5.10) by finite differencing Q (see Section 5.5.2):

```
ll.grad <- function(theta,vol,age,eps=1e-4) {
  q0 <- Q(theta,theta,vol,age)
  n <- length(theta); g <- rep(0,n)
  for (i in 1:n) {
    th <- theta; th[i] <- th[i] + eps
    g[i] <- (Q(th,theta,vol,age)-q0)/eps
  }
  g
}
```

Given `ll.grad` we do not really need (5.11), but can simply use finite differences of `ll.grad` to get the approximate Hessian of the log likelihood:

```
ll.hess <- function(theta,vol,age,eps=1e-4) {
  g0 <- ll.grad(theta,vol,age,eps)
  n <- length(theta); H <- matrix(0,n,n)
  for (i in 1:n) {
    th <- theta; th[i] <- th[i] + eps
    H[i,] <- (ll.grad(th,vol,age,eps)-g0)/eps
  }
  B <- solve(H)
  list(H=(H + t(H))/2,B=(B + t(B))/2)
}
```

Notice that the inverse Hessian is also computed, and both Hessian and inverse are made symmetric before returning. A Newton loop is then straightforward to implement: its only nonstandard feature is that step-length control must now be based on ensuring that, at the step end, the derivative in the direction of the step is not negative (see the end of Section 5.1.1).

```
for (i in 1:30) {
  g <- ll.grad(th,uv$vol,uv$age)
  B <- ll.hess(th,uv$vol,uv$age)$B
  eb <- eigen(B)
  if (max(eb$values)>0) { ## force neg def.
    d <- -abs(eb$values)
    B <- eb$vectors%*%(d*t(eb$vectors))
  }
  step <- -B%*%g; step <- step/max(abs(step))
  while(sum(step*ll.grad(th+step,uv$vol,uv$age))<0) {
    step <- step/2 }
  th <- th + step
  cat(th,mean(abs(g)),"\n")
  if (max(abs(g))<1e-4) break
}
```

Starting from the basic EM parameter estimates after 10 iterations, this loop converges in 12 further iterations. By contrast, after 30 steps of the basic EM iteration, some components of the log-likelihood gradient vector still had magnitude greater than 1. The step-length limitation, so that the maximum step component is of size 1, ensures that no ludicrously large steps can cause numerical problems in evaluating Q. Parameter estimates (log scale) and standard errors are as follows:

```
> th;diag(-ll.hess(th,uv$vol,uv$age)$B)^.5
[1] -3.39180 -0.36804 -1.69383 0.18609 -1.50680 -1.35400
[1] 0.712693 0.188591 0.168546 0.039671 0.188626 0.257298
```

For practical purposes these results are indistinguishable from the first-order Laplace approximation results. So in this case the extra effort of using higher order approximations to the log-likelihood is of little benefit, other than to confirm the results of the first order Laplace approximation. Actually, given that the difference between the MLE and the true parameters is typically $O(n^{-1/2})$ for the exact likelihood, it is not really surprising that we often see no substantial improvement in using an $O(n^{-2})$ approximation in place of an $O(n^{-1})$ approximation when computing the MLE.

Some model checking plots are shown in Figure 5.7. The random effects appear somewhat heavy tailed relative to normal: perhaps they should be modelled as t distributed. Otherwise the model appears fairly convincing, with much of the variability explained by urchin-to-urchin growth rate variability. The main practical objection is that the residual error is still a little high to be explicable as measurement error. The way to make further progress on this might be to seriously estimate the measurement error by separate calibration measurements and to include this measured measurement error in the model specification.

5.5 Computer differentiation

The preceding sections rely on a great deal of differentiation. If carried out by hand this can rapidly become tedious, and anyone whose sense of self worth is not reliant on carefully performing enormously lengthy routine calculations will quickly find themselves looking for automated alternatives. There are three possibilities:

1. Use a computer algebra system, such as Mathematica, Maple or Maxima[7] to help with the differentiation. This works well for relatively simple models, although the results often require a certain amount of 'hand simplification' before use.
2. Approximate the derivatives using finite differences. This is always possible, but is less accurate than the other methods.

[7] Maxima is free software.

3. Use *automatic differentiation* (AD), which computes numerically exact derivatives directly from the computer code implementing the function to be differentiated, by automatic application of the chain rule. Relative to approach 1, this is feasible for much more complicated situations.

5.5.1 Computer algebra

A general discussion of computer symbolic algebra is beyond the scope of this chapter. However, it is worth illustrating the basic symbolic differentiation available in R function `D`, which will symbolically differentiate R expressions with respect to single variables. As an example, consider differentiating $g(a,x) = \{\sin(ax)x^2\}^{-1}$ w.r.t. x:

```
> dx <- D(expression(1/(sin(a*x)*x^2)),"x"); dx
-((cos(a * x) * a * x^2 + sin(a * x) * (2 * x))/
  (sin(a * x) * x^2)^2)
```

The expression defined by the first argument is differentiated by the variable identified in the second argument (a character string). A '`call`' is returned, which can in turn be differentiated by `D`. For example, let us evaluate $\partial^2 g/\partial a \partial x$:

```
> D(dx,"a")
-(((cos(a * x) - sin(a * x) * x * a) * x^2 + cos(a * x)
  * x * (2 * x))/(sin(a * x) * x^2)^2 - (cos(a * x) * a
  * x^2 + sin(a * x) * (2 * x)) * (2 * (cos(a * x) * x
  * x^2 * (sin(a * x) * x^2)))/((sin(a * x) * x^2)^2)^2)
```

This result would clearly benefit from some simplification.

5.5.2 Finite differences

Consider differentiating a sufficiently smooth function $f(\mathbf{x})$ with respect to the elements of its vector argument $\mathbf{x}$. f might be something simple like $\sin(x)$ or something complicated like the mean global temperature predicted by an atmospheric global circulation model, given an atmospheric composition, forcing conditions and so on. A natural way to approximate the derivatives is to use the *finite difference* (FD) approximation:

$$\frac{\partial f}{\partial x_i} \simeq \frac{f(\mathbf{x} + \Delta \mathbf{e}_i) - f(\mathbf{x})}{\Delta}, \tag{5.15}$$

where Δ is a small constant and $\mathbf{e}_i$ is a vector of the same dimension as $\mathbf{x}$, with zeroes for each element except the ith, which is 1. How big should Δ be? As small as possible, right? Wrong. The difficulty is that computers only store real numbers to finite precision (usually equivalent to about 16 places of decimals for a 64-bit `double` precision floating point number). This means that if Δ is too small, there is a danger that the computed values of $f(\mathbf{x} + \Delta\mathbf{e}_i)$ and $f(\mathbf{x})$, will be identical and (5.15) will be in error by 100%. Even in less extreme situations, almost all precision can be lost. The following code snippet illustrates the issue:

```
> a <- 1e16; b <- a + pi
> b-a; pi
[1] 4
[1] 3.141593
```

Clearly the exact value of $b-a$ should be π, not 4, but π is a tiny proportion of 10^{16}. Hence when storing $10^{16} + \pi$ with about 16 places of decimals, we lose all the information from 3.141593 after the decimal point.[8] Such loss of precision is known as *cancellation error*.

It is possible to obtain a bound on the cancellation error involved in (5.15). Suppose that we can calculate f to one part in ϵ^{-1} (the best we can hope for here is that ϵ is the machine precision). Now let L_f be an upper bound on the magnitude of f, and denote the computed value of f by $\texttt{comp}(f)$. We have $|\texttt{comp}\{f(\mathbf{x}+\Delta\mathbf{e}_i)\} - f(\mathbf{x}+\Delta\mathbf{e}_i)| \leq \epsilon L_f$ and $|\texttt{comp}\{f(\mathbf{x})\} - f(\mathbf{x})| \leq \epsilon L_f$ which combine to imply that

$$\left| \frac{\texttt{comp}\{f(\mathbf{x}+\Delta\mathbf{e}_i) - f(\mathbf{x})\}}{\Delta} - \frac{f(\mathbf{x}+\Delta\mathbf{e}_i) - f(\mathbf{x})}{\Delta} \right| \leq \frac{2\epsilon L_f}{\Delta}.$$

So the right hand side is an upper bound on the cancellation error resulting from differencing two very similar quantities in finite precision arithmetic.

The cancellation error bound implies that we would like Δ to be as large as possible, but that would cause (5.15) to deteriorate as an approximation. To investigate this we need a bound on the error in (5.15) that occurs even if all the components on its right hand side are computed exactly. In a slightly sloppy notation, Taylor's theorem tells us that

$$f(\mathbf{x}+\Delta\mathbf{e}_i) = f(\mathbf{x}) + \nabla f(\mathbf{x})^{\mathrm{T}}\mathbf{e}_i\Delta + \frac{1}{2}\Delta^2\mathbf{e}_i^{\mathrm{T}}\nabla^2 f\mathbf{e}_i.$$

Rearranging while noting that $\nabla f(\mathbf{x})^{\mathrm{T}}\mathbf{e}_i = \partial f/\partial x_i$ we have

$$\frac{f(\mathbf{x}+\Delta\mathbf{e}_i) - f(\mathbf{x})}{\Delta} - \frac{\partial f}{\partial x_i} = \frac{1}{2}\Delta\mathbf{e}_i^{\mathrm{T}}\nabla^2 f\mathbf{e}_i.$$

[8] `b-a` is 4 rather than 3 as a result of representing numbers using binary, rather than decimal.

Now suppose that L is an upper bound on the magnitude of $\mathbf{e}_i^{\mathrm{T}} \nabla^2 f \mathbf{e}_i = \partial^2 f / \partial x_i^2$. It follows that

$$\left| \frac{f(\mathbf{x} + \Delta \mathbf{e}_i) - f(\mathbf{x})}{\Delta} - \frac{\partial f}{\partial x_i} \right| \leq \frac{L\Delta}{2}.$$

That is to say, we have an upper bound on the finite difference *truncation*[9] error.

So we want Δ to be as small as possible to minimise truncation error, and as large as possible to minimise cancellation error. Given that the total error is bounded as follows,

$$\text{err.fd} \leq \frac{L\Delta}{2} + \frac{2\epsilon L_f}{\Delta},$$

it makes sense to choose Δ to minimise the bound. That is, we should choose

$$\Delta \approx \sqrt{\frac{4\epsilon L_f}{L}}.$$

If the typical sizes of f and its second derivatives are similar, then

$$\Delta \approx \sqrt{\epsilon}$$

will not be too far from optimal. This is why the square root of the machine precision is often used as the finite difference interval. If $L_f \not\approx L$ or f is not calculable to a relative accuracy that is a small multiple of the machine precision, then consult §8.6 of Gill et al. (1981).

Other FD formulae

The finite difference approach just considered is *forward differencing*. Centred differences are more accurate, but more costly:

$$\frac{\partial f}{\partial x_i} \simeq \frac{f(\mathbf{x} + \Delta \mathbf{e}_i) - f(\mathbf{x} - \Delta \mathbf{e}_i)}{2\Delta}.$$

In the well-scaled case $\Delta \approx \epsilon^{1/3}$ is about right.

Higher order derivatives can also be useful. For example,

$$\frac{\partial^2 f}{\partial x_i \partial x_j} \simeq \frac{f(\mathbf{x} + \Delta \mathbf{e}_i + \Delta \mathbf{e}_j) - f(\mathbf{x} + \Delta \mathbf{e}_i) - f(\mathbf{x} + \Delta \mathbf{e}_j) + f(\mathbf{x})}{\Delta^2},$$

[9] So called because it is the error associated with *truncating* the Taylor series approximation to the function.

which in the well-scaled case will be most accurate for $\Delta \approx \epsilon^{1/4}$. Obviously if exact first derivatives are available it would be preferable to difference those.

5.5.3 Automatic differentiation

Automatic differentiation (AD) operates by differentiating a function based directly on the computer code that evaluates the function. There are several approaches, but the most elegant use the features of *object-oriented* programming languages to achieve the desired end. The key feature of an object-oriented language, from the AD perspective, is that every data structure, or `object`, in such a language has a *class* and the meaning of operators such as +, -, *, etc. depends on the class of the objects to which they are applied. Similarly the action of a function depends on the class of its arguments. See Section 3.6.

Suppose then, that we would like to differentiate

$$f(x_1, x_2, x_3) = \{x_1 x_2 \sin(x_3) + e^{x_1 x_2}\} / x_3$$

w.r.t. its real arguments x_1, x_2 and x_3.[10] In R the code

```
(x1*x2*sin(x3)+ exp(x1*x2))/x3
```

would evaluate the function, if `x1`, `x2` and `x3` were initialised to be floating point numbers.

Now define a new type of object of class `"ad"` that has a value (a floating point number) and a `"grad"` attribute. In the current case this `"grad"` attribute will be a 3-vector containing the derivatives of the value w.r.t. `x1`, `x2` and `x3`. We can now define versions of the arithmetic operators and mathematical functions that will return class `"ad"` results with the correct value and `"grad"` attribute, whenever they are used in an expression.

Here is an R function to create and initialise a simple class `"ad"` object:

```
ad <- function(x,diff = c(1,1)) {
## create class "ad" object. diff[1] is length of grad
## diff[2] is element of grad to set to 1.
  grad <- rep(0,diff[1])
  if (diff[2]>0 && diff[2]<=diff[1]) grad[diff[2]] <- 1
  attr(x,"grad") <- grad
  class(x) <- "ad"
  x
}
```

[10] This example function is taken from Nocedal and Wright (2006).

Here it is in use, initialising `x1` to 1, giving it a three dimensional `"grad"` attribute, and setting the first element of `grad` to 1, since $\partial x_1/\partial x_1 = 1$:

```
> x1 <- ad(1,c(3,1))
> x1
[1] 1
attr(,"grad")
[1] 1 0 0
attr(,"class")
[1] "ad"
```

Now the interesting part. We define versions of mathematical functions and operators that are specific to class `"ad"` objects and correctly propagate derivatives alongside values. Here is a `sin` function for class `"ad"`:

```
sin.ad <- function(a) {
  grad.a <- attr(a,"grad")
  a <- as.numeric(a) ## avoid infinite recursion!
  d <- sin(a)
  attr(d,"grad") <- cos(a) * grad.a ## chain rule
  class(d) <- "ad"
  d
}
```

Here is what happens when it is applied to `x1`:

```
> sin(x1)
[1] 0.841471
attr(,"grad")
[1] 0.5403023 0.0000000 0.0000000
attr(,"class")
[1] "ad"
```

So the value of the result is $\sin(x_1)$ and the first element of its `"grad"` contains the derivative of $\sin(x_1)$ w.r.t. x_1 evaluated at $x_1 = 1$.

Operators can also be *overloaded* in this way. For example, here is the multiplication operator for class `"ad"`:

```
"*.ad" <- function(a,b) { ## ad multiplication
  grad.a <- attr(a,"grad")
  grad.b <- attr(b,"grad")
  a <- as.numeric(a)
  b <- as.numeric(b)
  d <- a*b ## evaluation
  attr(d,"grad") <- a * grad.b + b * grad.a ## chain rule
  class(d) <- "ad"
  d
}
```

Continuing in the same way we can provide a complete library of mathematical functions and operators for the `"ad"` class. Given such a library, we can obtain the derivatives of a function directly from the code that would simply evaluate it, given ordinary floating point arguments. For example, here is some code evaluating the example function:

```
> x1 <- 1; x2 <- 2; x3 <- pi/2
> (x1*x2*sin(x3)+ exp(x1*x2))/x3
[1] 5.977259
```

and here is the same code with the arguments replaced by `"ad"` objects:

```
> x1 <- ad(1,c(3,1))
> x2 <- ad(2,c(3,2))
> x3 <- ad(pi/2,c(3,3))
> (x1*x2*sin(x3)+ exp(x1*x2))/x3
[1] 5.977259
attr(,"grad")
[1] 10.681278 5.340639 -3.805241
attr(,"class")
[1] "ad"
```

You can check that these results are correct (actually to machine accuracy).

This simple propagation of derivatives alongside the evaluation of a function is known as *forward* mode auto-differentiation. R is not the best language in which to try to do this, and if you need AD for complex models it is often better to use existing software libraries in C++, for example, which have done all the function and operator rewriting for you.

The `deriv` *function in R*

For functions that are not overly complex, R function `deriv` implements forward mode AD using a 'source translation', rather than an operator overloading method. The expression to be differentiated is supplied as an R expression or one-sided formula, along with a character vector specifying the variables with respect to which to differentiate. Repeating the preceding example we have:

```
> f <- expression((x1*x2*sin(x3)+ exp(x1*x2))/x3)
> g <- deriv(f,c("x1","x2","x3"),
+             function.arg=c("x1","x2","x3"))
> g(1,2,pi/2)
[1] 5.977259
attr(,"gradient")
           x1       x2        x3
[1,] 10.68128 5.340639 -3.805241
```

The argument `function.arg` tells `deriv` that we want a function (rather than an expression) to be returned and what its arguments should be. There is a further argument `hessian`, which if `TRUE` causes second derivatives to be computed along with the gradients.

A caveat

For AD to work, it is not sufficient that the function being evaluated has properly defined derivatives at the evaluated function value. It requires that every function/operator used in the evaluation has properly defined derivatives at its evaluated argument(s). This can create a problem with code that executes conditionally on the value of some variable. For example, the Box-Cox transformation of a positive datum y is

$$B(y;\lambda) = \begin{cases} (y^\lambda - 1)/\lambda & \lambda \neq 0 \\ \log(y) & \lambda = 0 \end{cases}.$$

If you code this up in the obvious way, then AD will never get the derivative of B w.r.t. λ right if $\lambda = 0$.

Reverse-mode AD

If you require many derivatives of a scalar valued function, then forward-mode AD will have a theoretical computational cost similar to finite differencing, because at least as many operations are required for *each* derivative as are required for function evaluation. In reality the overheads associated with operator overloading make AD more expensive and alternative strategies also carry overheads. Of course, the benefit of AD is higher accuracy, and in many applications the cost is not critical.

An alternative with the potential for *big* computational savings is *reverse*-mode AD. Again concentrate on the Nocedal and Wright (2006) example:

$$f(x_1, x_2, x_3) = \left\{x_1 x_2 \sin(x_3) + e^{x_1 x_2}\right\}/x_3.$$

Any computer evaluating f must break the computation down into a sequence of elementary operations on one or two floating point numbers. This can be thought of as a *computational graph*:

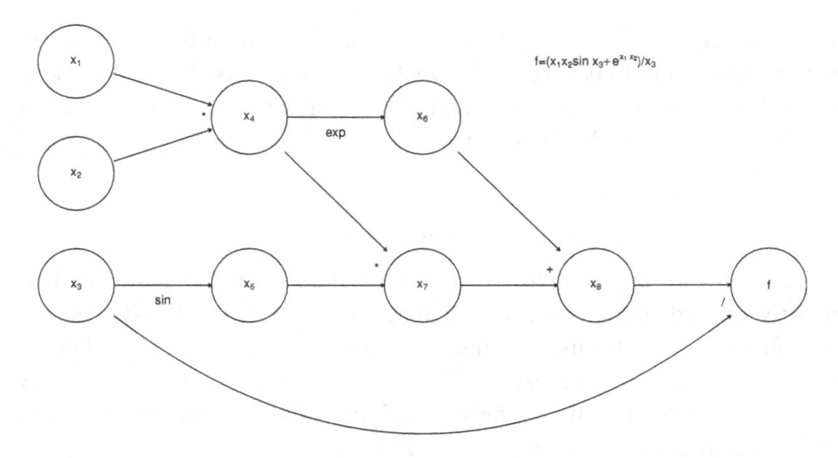

where the nodes x_4 to x_8 are the intermediate quantities that will have to be produced en route from the input values x_1 to x_3 to the final answer, f. The arrows run from *parent* nodes to *child* nodes. No child can be evaluated until all its parents have been evaluated. Simple left-to-right evaluation of this graph results in this:

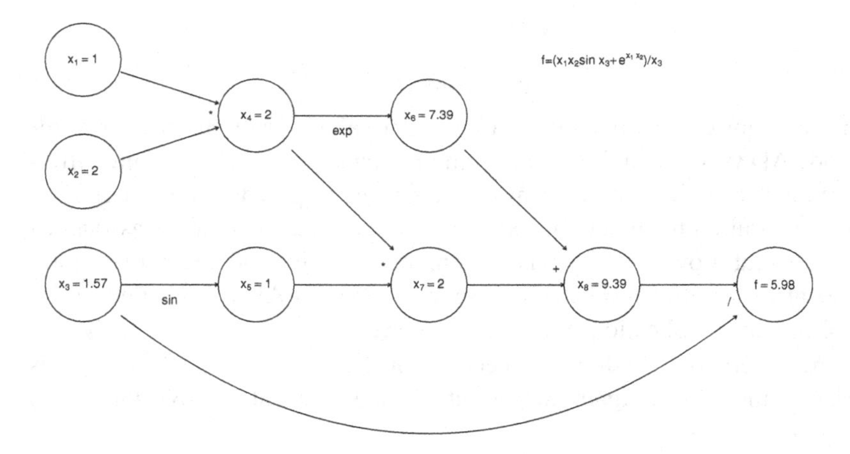

Now, forward-mode AD carries derivatives forward through the graph, alongside values. For example, the derivative of a node with respect to input variable x_1 is computed using

$$\frac{\partial x_k}{\partial x_1} = \sum_{j \text{ parent of } k} \frac{\partial x_k}{\partial x_j} \frac{\partial x_j}{\partial x_1},$$

the right hand side being evaluated by overloaded functions and operators, in the object oriented approach. The following illustrates this process, just for the derivative w.r.t. x_1:

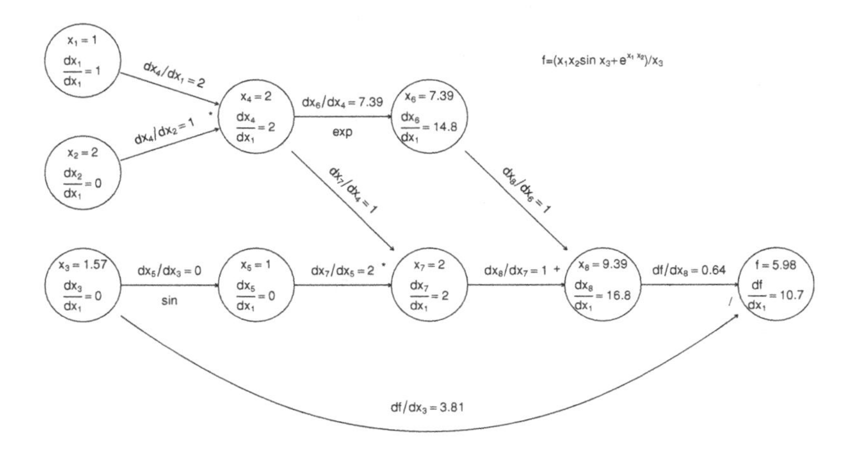

Again computation runs left to right, with evaluation of a node only possible once all parent values are known.

If we require derivatives w.r.t. several input variables, then each node will have to evaluate derivatives w.r.t. each of these variables, and this becomes expensive (in the previous graph, each node would contain multiple evaluated derivatives). Reverse mode therefore does something ingenious. It first executes a *forward sweep* through the graph, evaluating the function and all the derivatives of nodes w.r.t. their parents, as follows:

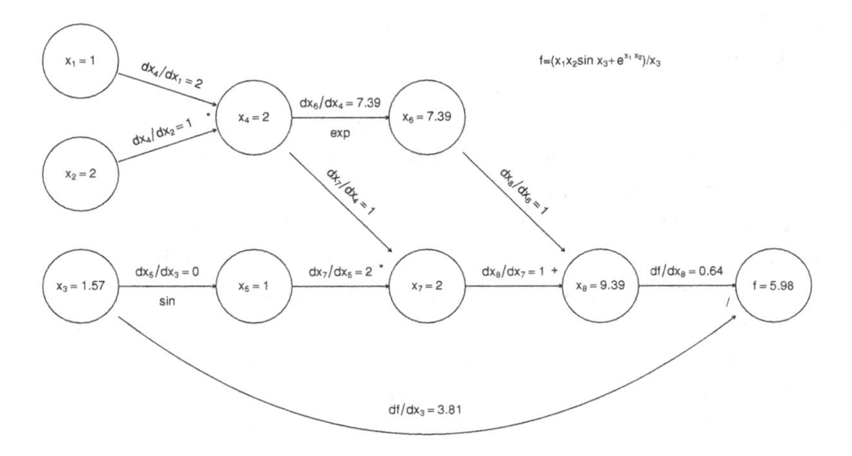

The *reverse* sweep then works backwards from the terminal node, for which $\partial f / \partial f = 1$, evaluating the derivative of f w.r.t. each node using

$$\frac{\partial f}{\partial x_k} = \sum_{j \text{ is child of k}} \frac{\partial x_j}{\partial x_k} \frac{\partial f}{\partial x_j}.$$

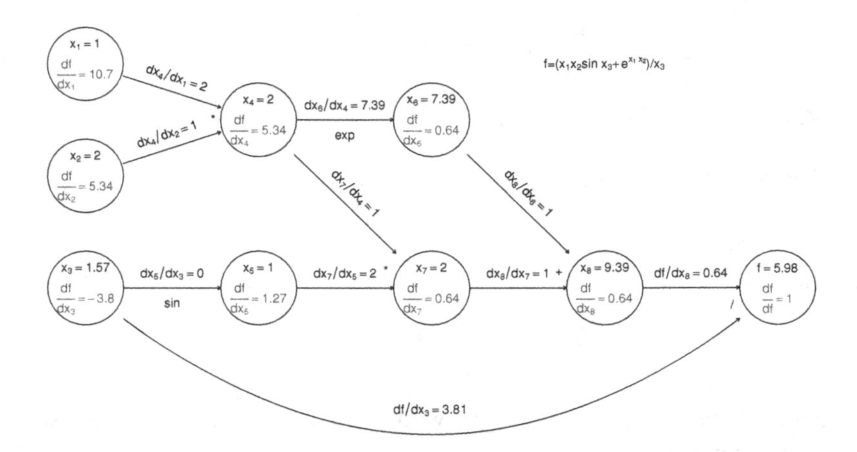

The derivatives in grey are those calculated on the reverse sweep. The point here is that there is only one derivative to be evaluated at each node, but in the end we know the derivative of f w.r.t. every input variable. Reverse-mode AD can therefore save a large number of operations relative to finite differencing or forward-mode AD. Once again, general-purpose AD libraries automate the process for you, so that all you need to be able to write is the evaluation code.

Unfortunately, reverse-mode efficiency comes at a heavy price. In forward mode we could discard the values and derivatives associated with a node as soon as all its children were evaluated. In reverse mode the values of *all nodes* and the evaluated derivatives associated with *every connection* have to be stored during the forward sweep in order to be used in the reverse sweep. This is a heavy storage requirement. For example, if f involved the inversion of a 1000×1000 matrix then we would have to store some 2×10^9 intermediate node values plus a similar number of evaluated derivatives. That amounts to some 32 Gigabytes of storage before we even consider the requirements for storing the *structure* of the graph. Much research is concerned with hybrid AD strategies to simultaneously reduce the operation and memory costs. See Griewank and Walther (2008) for more.

Using AD to improve FD

When fitting complicated or computer intensive models AD may be too expensive to use for routine derivative calculation during optimisation. However, it can still provide a useful means for calibrating FD intervals. A 'typical' model run can be autodifferentiated and the finite difference intervals adjusted to achieve the closest match to the AD derivatives. As optimisation progresses, one or two further calibrations of the FD intervals can be carried out as necessary.

5.6 Looking at the objective function

Given the apparent generality of the preceding theory and methods, it is easy to assume that if you can evaluate the log likelihood (or other objective function), then it will be possible to optimise it and draw useful statistical conclusions from the results. This assumption is not always true, and it is prudent to produce plots to check that the objective is the well behaved function imagined.

A simple example emphasises the importance of these checks. Consider fitting an apparently innocuous dynamic model to a single time series by least squares/maximum likelihood. The model is

$$n_{t+1} = rn_t(1 - n_t/K), \quad t = 0, 1, 2, \ldots,$$

where r and K are parameters and we will assume that n_0 is known. Further suppose that we have observations $y_t = n_t + \epsilon_t$ where $\epsilon_t \underset{\text{i.i.d.}}{\sim} N(0, \sigma^2)$ and σ is known. Estimation of r and K by least squares (or maximum likelihood, in this case) requires minimisation of

$$f(r, K) = \sum_i \{y_i - n_i(r, K)\}^2$$

w.r.t. r and K. We should try to get a feel for the behaviour of f. To see how this can work, consider two simulated data examples. In each case I used $n_0 = 20$, $K = 50$ and $\sigma = 1$ for the simulations, but varied r between the cases.

- In the first instance data were simulated with $r = 2$. If we now pretend that we need to estimate r and K from such data, then we might look at some r-transects and some K-transects through f. This figure shows the raw data and an r-transect.

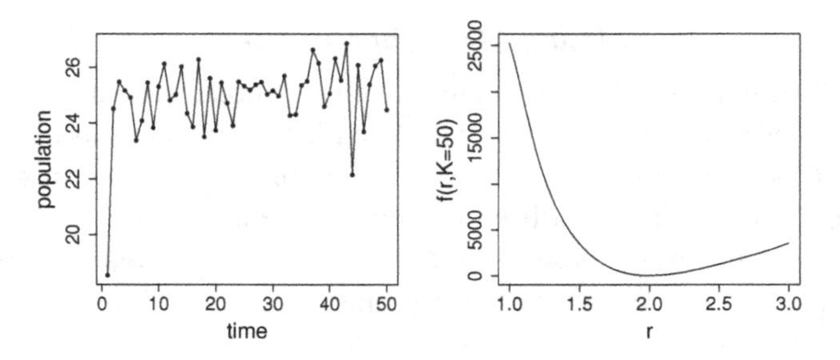

r-transects with other K values look equally innocuous, and K-transects also look benign over this r range. So in this case f appears to be a nice smooth function of r and K, and any half-decent optimisation method ought to be able to find the optimum.

- In the second case data were simulated with $r = 3.8$.

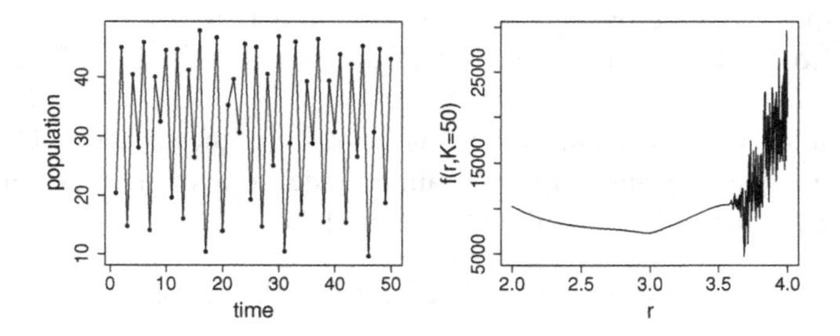

Now the objective has a minimum somewhere around 3.7, but it is surrounded by other local minima in a highly irregular region, so that locating the actual minima would be a rather taxing problem. In addition it is now unclear how we would go about quantifying uncertainty about the 'optimal' $\boldsymbol{\theta}$: it will certainly be of no use appealing to asymptotic likelihood arguments in this case.

In both of these examples, simple transects through the objective function provided useful information. In the first case everything seemed OK. In the second case we would need to think very carefully about the purpose of the optimisation, and about whether a reformulation of the basic problem might be needed. Notice how the behaviour of the objective was highly parameter dependent, something that emphasises the need to understand models quite well before trying to fit them. In this case the dynamic model, although very simple, can show a wide range of complicated behaviour, including chaos.

5.6.1 Objective function transects are a partial view

Plotting transects through the objective function is a good idea, but they can only give a limited and partial view when $\boldsymbol{\theta}$ is multidimensional. For example, the left hand plot, below, shows an x-transect through a function, $f(x, y)$, plotted on the right.

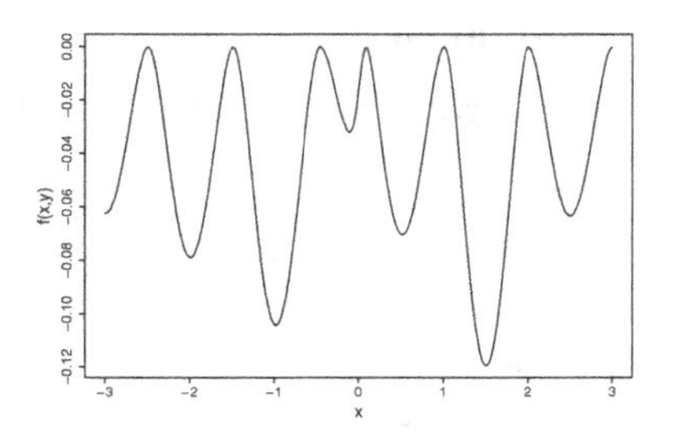

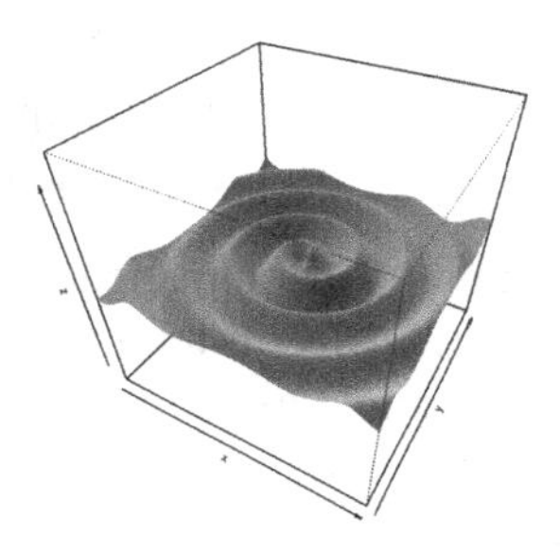

From the left-hand plot it appears that the function has many local minima and optimisation will be difficult. But in fact it has one local minimum, its global minimum. Head downhill from any point x, y and you will eventually reach the minimum, as the right-hand plot shows.

The opposite problem can also occur. Here are x and y transects through a second function $g(x, y)$:

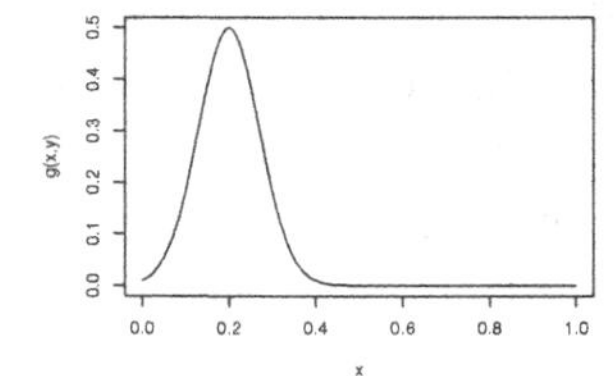

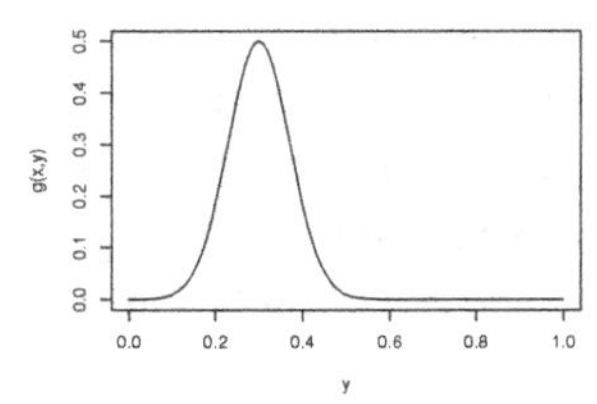

From these plots you would be tempted to conclude that g is well behaved and unimodal. The problem is that g actually looks like this:

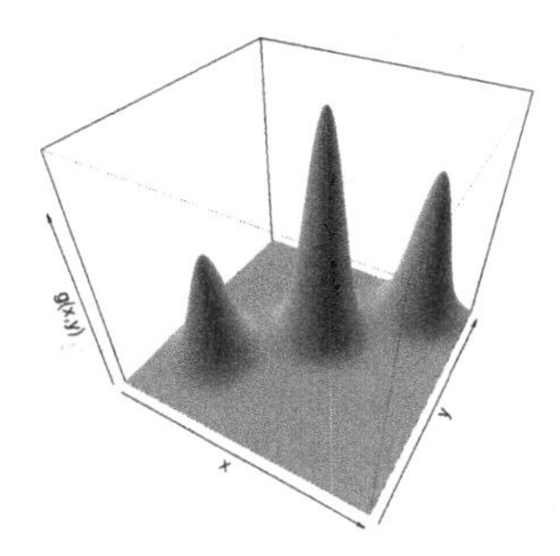

So, generally speaking, it is a good idea to plot transects through the objective function before optimisation, and to plot transects passing through the apparent optimum after optimisation. However, bear in mind that transects only give partial views.

5.7 Dealing with multimodality

There is no universal prescription for dealing with multimodality, but the following approaches may help:

- A common recommendation is to repeat optimisation a number of times from radically different starting values, perhaps randomly generated. This can help to indicate multimodality and to find the dominant mode.
- For relatively small-scale local optima, a *bootstrapping* approach can be helpful. Suppose we have a log likelihood, $l(\boldsymbol{\theta})$, based on data vector $\mathbf{y}$. Start with a parameter guess, $\boldsymbol{\theta}_0$, and iterate the following steps:

  1. Starting at $\boldsymbol{\theta}_0$, seek $\hat{\boldsymbol{\theta}} = \text{argmax}_\theta\, l(\boldsymbol{\theta})$, by numerical optimisation.
  2. Resample your data with replacement to produce a resampled data vector $\mathbf{y}^*$ and corresponding log-likelihood function $l^*(\boldsymbol{\theta})$.
  3. Starting at $\hat{\boldsymbol{\theta}}$, seek $\boldsymbol{\theta}_0 = \text{argmax}_\theta\, l^*(\boldsymbol{\theta})$, by numerical optimisation.

  Any auxiliary data are resampled alongside $\mathbf{y}$ (so that auxilliary data stay with the datum to which they belong). The idea is that by randomly perturbing the objective, it may be possible to escape local optima. For greater perturbation, smaller re-samples can be used.
- It the objective appears pathologically multimodal, it is probably time to reformulate the question being addressed.

Exercises

5.1 Rosenbrock's function $f(x, z) = a(z - x^2)^2 + (b - x)^2$ is a classic test function for optimisation methods. Usually $a = 100$ and $b = 1$.

  a. Write a function `Rosenbrock` with vector arguments `x` and `z` and scalar arguments `a` and `b`, with default values of 10 and 1, respectively. Using `contour` and `outer`, produce a contour plot of f for $-1.5 \leq x \leq 1.5$ and $-0.5 \leq z \leq 1.5$.
  b. Write a modified version of `Rosenbrock`, suitable for optimisation using `optim`. Optimise `Rosenbrock` using `optim` with starting values $x = -1$, $z = 1$, and compare results using Nelder-Mead and BFGS.
  c. Repeat the optimisation using `nlm` and `nlminb`.

5.2 Write your own code to optimise Rosenbrock's function by Newton's method. Optionally use R's symbolic or automatic differentiation functions to obtain the required gradient and Hessian.

5.3 Write your own code to optimise Rosenbrock's function by BFGS.

5.4 Write functions suitable for use with `optim` to evaluate the negative log likelihood and its gradient for the cell count example in Section 5.1.1. Hence, find the MLE of δ using `optim`. Compute and compare 95% confidence intervals for δ based on the Hessian of the log likelihood and generalised likelihood ratio test inversion.

5.5 Write a function suitable for use with `optim` to evaluate the negative log likelihood of the AIDS cases model of Section 5.1.1 (use `dpois` with the `log=TRUE` option). Write a second function evaluating the negative log likelihood of an extended version of the model in which the dependence of the log case rate on time is quadratic, rather than linear. Use a generalised likelihood ratio test (GLRT) to compare the models, and also compute their AIC values. What aspect of the GLRT might be suspect here?

5.6 R package `MASS` contains a dataframe `geyser` where `geyser$waiting` gives the waiting times between eruptions of the Old Faithful geyser in Yellowstone National Park. A possible model is that that the waiting times, t_i, are independently drawn from a mixture of two normal distributions, with p.d.f.

$$f(t_i) = \frac{\phi}{\sqrt{2\pi}\sigma_1} e^{-\frac{1}{2\sigma_1^2}(t_i-\mu_1)^2} + \frac{1-\phi}{\sqrt{2\pi}\sigma_2} e^{-\frac{1}{2\sigma_2^2}(t_i-\mu_2)^2},$$

where parameter ϕ is bounded between 0 and 1. Find MLEs for the parameters and test whether $p = 0.5$. Are there any theoretical caveats on this analysis? Produce appropriate model checking plots (*not* 'residual' plots!).

5.7 R package `faraway` contains data from an experiment on balance in human subjects. There were 40 subjects, and the experiment was repeated a number of times on each, standing on two different surfaces under three different degrees of restriction of vision. Sex, age, height and weight of the subjects were recorded. The following code loads the data, creates a variable indicating whether subjects were judged fully `stable` (1) or not (0), and converts the subject identifiers to a `factor` variable:

```
library(faraway)
ctsib$stable <- ifelse(ctsib$CTSIB==1,1,0)
ctsib$Subject <- factor(ctsib$Subject)
```

Interest lies is in explaining stability in terms of the other variables. A possible model for the data involves a vector of subject-specific random effects $\mathbf{b}$, and is as follows:

$$\texttt{stable}_i|\mathbf{b} \sim \text{Bernoulli}(\mu_i) \quad \mu_i = e^{\eta_i}/(1+e_i^{\eta}),$$

where if measurement i if for subject j, of sex k, on surface m, and with vision restriction q, then

$$\eta_i = \alpha + \gamma_k + \delta_m + \phi_q + \tilde{\beta}\texttt{Height}_i + b_j,$$

where $b_j \sim N(0, \sigma_b^2)$ (independent). The subject-specific random effects are important, because we would expect subject-to-subject variability in their balancing ability. More generally we can write the model for η_i in vector matrix form as

$$\boldsymbol{\eta} = \mathbf{X}\boldsymbol{\beta} + \mathbf{Z}\mathbf{b}, \quad \mathbf{b} \sim N(\mathbf{0}, \mathbf{I}\sigma_b^2).$$

$\mathbf{X}$ contains a column of `Height` data and columns of zeroes and ones identifying measurements to particular groups (`Sex`, `Vision`, etc.). $\mathbf{Z}$ contains zeroes and ones in order to pick out the correct b_j for each data row. The following code creates suitable matrices:

```
X <- model.matrix(~ Sex+Height+Surface+Vision,ctsib)
Z <- model.matrix(~ Subject-1,ctsib)
```

a. Write an R function to evaluate the joint probability/density of `stable` and the subject-specific random effects, along with its gradient and the leading diagonal of its Hessian w.r.t. $\mathbf{b}$. Only the leading diagonal of the Hessian is required, as it turns out to be a diagonal matrix.
b. Write an R function to evaluate the negative log likelihood of the model parameters, integrating out $\mathbf{b}$ by Laplace approximation.
c. Fit the model using `optim` to find the MLE.

6

Bayesian computation

Recall that the Bayesian approach to statistics views the model parameters, $\boldsymbol{\theta}$, as random variables with *prior* p.d.f., $f(\boldsymbol{\theta})$, and then answers the basic questions of statistical inference using the *posterior* p.d.f.

$$f(\boldsymbol{\theta}|\mathbf{y}) = \frac{f(\mathbf{y}|\boldsymbol{\theta})f(\boldsymbol{\theta})}{f(\mathbf{y})},$$

(see Section 2.5). The principal practical challenges are that

$$f(\mathbf{y}) = \int f(\mathbf{y}|\boldsymbol{\theta})f(\boldsymbol{\theta})d\boldsymbol{\theta} \tag{6.1}$$

is usually intractable for interesting models and that it is usually equally intractable integrals of $f(\boldsymbol{\theta}|\mathbf{y})$ that are of direct interest. There are then two main strategies for making progress: either approximate the required integrals or find a way of simulating from $f(\boldsymbol{\theta}|\mathbf{y})$ without requiring such integrals. The latter strategy is based on the fact that, for many statistical purposes, the ability to simulate from a density is as good as being able to evaluate the density, and sometimes better. Hybrid strategies are also useful. For much more on the topics covered here see Gamerman and Lopes (2006), Robert and Casella (2009), Gelman et al. (2013) and, at a more advanced level, Robert (2007).

6.1 Approximating the integrals

One possibility is to evaluate the normalising constant (6.1), and other interesting integrals, using Laplace approximation. For example, integrate out $\boldsymbol{\theta}$ in exactly the same way as $\mathbf{b}$ was integrated out in Section 5.3.1. This relies on the integrand having only one important mode and, in the context of (6.1), is unlikely to result in an exactly proper posterior $f(\boldsymbol{\theta}|\mathbf{y})$.

Another approximation is based on recognising, from (6.1), that $f(\mathbf{y}) = E_\theta\{f(\mathbf{y}|\boldsymbol{\theta})\}$, and to approximate this expectation using the simulation technique known as *importance sampling*. The idea is simple. Suppose we want to estimate $\alpha = E_f\{\phi(\mathbf{X})\}$ where $\mathbf{X} \sim f(\mathbf{x})$. An obvious unbiased estimator is obtained by simulating n deviates, $\mathbf{x}_i$, from $f(\mathbf{x})$ and setting $\hat{\alpha}' = \sum_i \phi(\mathbf{x}_i)/n$. The problem is that $\phi(\mathbf{x}_i)$ may be very close to zero for many of the $\mathbf{x}_i$, so that our estimate is really based only on the few points for which ϕ was non-negligible, making the approximation inaccurate and highly variable. If we have access to a p.d.f. $g(\mathbf{z})$ that has high probability where $\phi(\mathbf{z})$ is high, and low probability otherwise, then we could use the fact that $\alpha = E_f\{\phi(\mathbf{X})\} = E_g\{\phi(\mathbf{Z})f(\mathbf{Z})/g(\mathbf{Z})\}$ to obtain the alternative unbiased estimator:

$$\tilde{\alpha} = \frac{1}{n}\sum_{i=1}^{n} \phi(\mathbf{z}_i)f(\mathbf{z}_i)/g(\mathbf{z}_i) \text{ where } \mathbf{z}_i \sim g(\mathbf{z}).$$

This importance sampling estimator tends to improve on the naive version, by placing the $\mathbf{z}_i$ in better locations w.r.t. ϕ. The $f(\mathbf{z}_i)/g(\mathbf{z}_i)$ are known as *importance weights*. A problem in Bayesian analysis is that $f(\mathbf{z}_i)$ is often an un-normalised density, so it is necessary to normalise the importance weights, leading to the modified importance sampling estimator,

$$\hat{\alpha} = \frac{\sum_{i=1}^{n} \phi(\mathbf{z}_i)f(\mathbf{z}_i)/g(\mathbf{z}_i)}{\sum_{i=1}^{n} f(\mathbf{z}_i)/g(\mathbf{z}_i)} \text{ where } \mathbf{z}_i \sim g(\mathbf{z}),$$

which can also be used directly for other integrals of more immediate interest than (6.1).

In the context of (6.1), it can be attractive to use the p.d.f. of

$$N(\hat{\boldsymbol{\theta}}, \{-\nabla_\theta^2 \log f(\mathbf{y}, \hat{\boldsymbol{\theta}})\}^{-1}) \tag{6.2}$$

as $g(\boldsymbol{\theta})$, where $\hat{\boldsymbol{\theta}}$ is the maximiser of $f(\mathbf{y}, \boldsymbol{\theta})$. This is motivated by the Laplace approximation. Alternatively, to reduce the risk of extreme weights in the tails of the distribution, $t_k(\hat{\boldsymbol{\theta}}, \{-\nabla_\theta^2 \log f(\mathbf{y}, \hat{\boldsymbol{\theta}})\}^{-1})$ can be used for g, with k set to a small integer (see Section 1.6.1). Similar proposals can be constructed for other integrals of interest.

Usually the quantities required from the posterior are expectations of functions of $\boldsymbol{\theta}$ according to the posterior distribution. That is, we require integrals of the form

$$\int \phi(\boldsymbol{\theta})f(\boldsymbol{\theta}|\mathbf{y})d\boldsymbol{\theta} = \frac{\int \phi(\boldsymbol{\theta})f(\mathbf{y}|\boldsymbol{\theta})f(\boldsymbol{\theta})d\boldsymbol{\theta}}{\int f(\mathbf{y}|\boldsymbol{\theta})f(\boldsymbol{\theta})d\boldsymbol{\theta}},$$

and these can be estimated by direct application of (5.14) in Section 5.3.2.

6.2 Markov chain Monte Carlo

The approximation methods just described are useful when we know that the posterior distribution has a relatively simple form, preferably with a single mode. However, these assumptions are doubtful in many cases where the Bayesian approach is appealing, and more general methods are then required. The key is to use the fact that

$$f(\boldsymbol{\theta}|\mathbf{y}) \propto f(\boldsymbol{\theta}, \mathbf{y}) \quad (= f(\mathbf{y}|\boldsymbol{\theta})f(\boldsymbol{\theta})),$$

to devise methods for simulating from $f(\boldsymbol{\theta}|\mathbf{y})$, which only require evaluation of $f(\boldsymbol{\theta}, \mathbf{y})$ with the observed data values plugged in for $\mathbf{y}$. The resulting Markov chain Monte Carlo (MCMC) methods simulate (correlated) samples from the distribution of the model unknowns (parameters and any random effects), given the data. Based on the unknowns at one step, a new set of unknowns is generated in such a way that the stable distribution of the resulting Markov chain is the distribution of interest. The development of MCMC methods relies on being able to generate apparently random numbers by computer: Appendix C discusses the extent to which this is possible.

6.2.1 Markov chains

To use MCMC we do not require much theoretical background on Markov chains, but some basic concepts are needed. A sequence of random vectors, $\mathbf{X}_1, \mathbf{X}_2, \mathbf{X}_3, \ldots$, constitutes a *Markov chain* if, for any j,

$$f(\mathbf{x}_j|\mathbf{x}_{j-1}, \mathbf{x}_{j-2}, \ldots, \mathbf{x}_1) = f(\mathbf{x}_j|\mathbf{x}_{j-1}).$$

For notational convenience let us rewrite the density of $\mathbf{x}_j$ given $\mathbf{x}_{j-1}$ as $P(\mathbf{x}_j|\mathbf{x}_{j-1})$, the *transition kernel* of the Markov chain. If there exists a density f_x such that

$$f_x(\mathbf{x}_j) = \int P(\mathbf{x}_j|\mathbf{x}_{j-1}) f_x(\mathbf{x}_{j-1}) d\mathbf{x}_{j-1}$$

(where f_x denotes the same density on both sides), then this is the *stationary distribution* of the chain. Existence of a stationary distribution depends on P being *irreducible*, meaning that wherever we start the chain, there is a positive probability of visiting all possible values of $\mathbf{X}$. If the chain is also *recurrent*, meaning that if its length tends to infinity it will revisit any non-negligible set of values an infinite number of times, then its stationary distribution is also its *limiting distribution*. This means that the chain can be started from any possible value of $\mathbf{X}$, and its marginal distribution will eventually converge on f_x. In consequence as the simulation length, J, tends to infinity,

$$\frac{1}{J}\sum_{j=1}^{J}\phi(\mathbf{X}_j) \to E_{f_x}\{\phi(\mathbf{X})\}$$

(also known as *ergodicity*). This extension of the law of large numbers (see Section 1.10.2) to this particular sort of correlated sequence is what makes MCMC methods useful, so the methods discussed in this chapter will be set up to produce chains with the required properties.

6.2.2 Reversibility

Now we turn to the issue of constructing Markov chains to generate sequences $\boldsymbol{\theta}_1, \boldsymbol{\theta}_2, \ldots$ from $f(\boldsymbol{\theta}|\mathbf{y})$. An MCMC scheme will generate samples from $f(\boldsymbol{\theta}|\mathbf{y})$, if it satisfies the *detailed balance* condition (also termed *reversibility*). Let $P(\boldsymbol{\theta}_i|\boldsymbol{\theta}_j)$ be the p.d.f. of $\boldsymbol{\theta}_i$ given $\boldsymbol{\theta}_j$, according to the chain. We require

$$P(\boldsymbol{\theta}_j|\boldsymbol{\theta}_{j-1})f(\boldsymbol{\theta}_{j-1}|\mathbf{y}) = P(\boldsymbol{\theta}_{j-1}|\boldsymbol{\theta}_j)f(\boldsymbol{\theta}_j|\mathbf{y}). \tag{6.3}$$

The left hand side of (6.3) is the joint p.d.f. of $\boldsymbol{\theta}_j, \boldsymbol{\theta}_{j-1}$ from the chain, if $\boldsymbol{\theta}_{j-1}$ is from $f(\boldsymbol{\theta}|\mathbf{y})$. Integrating w.r.t. $\boldsymbol{\theta}_{j-1}$ gives the corresponding marginal density of $\boldsymbol{\theta}_j$,

$$\begin{aligned}\int P(\boldsymbol{\theta}_j|\boldsymbol{\theta}_{j-1})f(\boldsymbol{\theta}_{j-1}|\mathbf{y})d\boldsymbol{\theta}_{j-1} &= \int P(\boldsymbol{\theta}_{j-1}|\boldsymbol{\theta}_j)f(\boldsymbol{\theta}_j|\mathbf{y})d\boldsymbol{\theta}_{j-1}\\ &= f(\boldsymbol{\theta}_j|\mathbf{y}).\end{aligned}$$

That is, given $\boldsymbol{\theta}_{j-1}$ from $f(\boldsymbol{\theta}|\mathbf{y})$, the chain generates $\boldsymbol{\theta}_j$ also from $f(\boldsymbol{\theta}|\mathbf{y})$ as result of (6.3). So provided that we start with a $\boldsymbol{\theta}_1$ that is not impossible according to $f(\boldsymbol{\theta}|\mathbf{y})$, then the chain will generate from the target distribution. How quickly it will converge to the high-probability region of $f(\boldsymbol{\theta}|\mathbf{y})$ is another matter.

6.2.3 Metropolis Hastings

The Metropolis-Hastings method constructs a chain with an appropriate P. It works as follows:

1. Pick a *proposal distribution* $q(\boldsymbol{\theta}_j|\boldsymbol{\theta}_{j-1})$ (e.g. a normal centred on $\boldsymbol{\theta}_{j-1}$). Then pick a value $\boldsymbol{\theta}_0$, set $j = 1$ and iterate steps 2 and 3:

2. Generate $\boldsymbol{\theta}'_j$ from $q(\boldsymbol{\theta}_j|\boldsymbol{\theta}_{j-1})$.
3. Set $\boldsymbol{\theta}_j = \boldsymbol{\theta}'_j$ with probability

$$\alpha = \min\left\{1, \frac{f(\mathbf{y}|\boldsymbol{\theta}'_j)f(\boldsymbol{\theta}'_j)q(\boldsymbol{\theta}_{j-1}|\boldsymbol{\theta}'_j)}{f(\mathbf{y}|\boldsymbol{\theta}_{j-1})f(\boldsymbol{\theta}_{j-1})q(\boldsymbol{\theta}'_j|\boldsymbol{\theta}_{j-1})}\right\}, \tag{6.4}$$

otherwise setting $\boldsymbol{\theta}_j = \boldsymbol{\theta}_{j-1}$. Increment j.

Note that the q terms cancel if q depends only on the magnitude of $\boldsymbol{\theta}_j - \boldsymbol{\theta}_{j-1}$ (e.g. if q is a normal centred on $\boldsymbol{\theta}_{j-1}$). The same goes for the prior densities, $f(\boldsymbol{\theta})$, if they are improper uniform. If both of these simplifications hold then we have $\alpha = \min\left\{1, L(\boldsymbol{\theta}'_j)/L(\boldsymbol{\theta}_{j-1})\right\}$, so that we are accepting or rejecting on the basis of the likelihood ratio.

An important consideration is that $\boldsymbol{\theta}_1$ may be very improbable so that the chain may take many iterations to reach the high-probability region of $f(\boldsymbol{\theta}|\mathbf{y})$. For this reason we usually need to discard a *burn-in period* consisting of the first few hundred or thousand $\boldsymbol{\theta}_j$ vectors simulated.

6.2.4 Why Metropolis Hastings works

As we saw in Section 6.2.2, the Metropolis-Hastings (MH) method will work if it satisfies detailed balance. It does, and proof is easy. To simplify notation let $\pi(\boldsymbol{\theta}) = f(\boldsymbol{\theta}|\mathbf{y}) \propto f(\mathbf{y}|\boldsymbol{\theta})f(\boldsymbol{\theta})$, so that the MH acceptance probability from $\boldsymbol{\theta}$ to $\boldsymbol{\theta}'$ is

$$\alpha(\boldsymbol{\theta}', \boldsymbol{\theta}) = \min\left\{1, \frac{\pi(\boldsymbol{\theta}')q(\boldsymbol{\theta}|\boldsymbol{\theta}')}{\pi(\boldsymbol{\theta})q(\boldsymbol{\theta}'|\boldsymbol{\theta})}\right\}.$$

We need to show that $\pi(\boldsymbol{\theta})P(\boldsymbol{\theta}'|\boldsymbol{\theta}) = \pi(\boldsymbol{\theta}')P(\boldsymbol{\theta}|\boldsymbol{\theta}')$. This is trivial if $\boldsymbol{\theta}' = \boldsymbol{\theta}$. Otherwise we know that $P(\boldsymbol{\theta}'|\boldsymbol{\theta}) = q(\boldsymbol{\theta}'|\boldsymbol{\theta})\alpha(\boldsymbol{\theta}', \boldsymbol{\theta})$, from which it follows that

$$\begin{aligned}\pi(\boldsymbol{\theta})P(\boldsymbol{\theta}'|\boldsymbol{\theta}) &= \pi(\boldsymbol{\theta})q(\boldsymbol{\theta}'|\boldsymbol{\theta})\min\left\{1, \frac{\pi(\boldsymbol{\theta}')q(\boldsymbol{\theta}|\boldsymbol{\theta}')}{\pi(\boldsymbol{\theta})q(\boldsymbol{\theta}'|\boldsymbol{\theta})}\right\} \\ &= \min\left\{\pi(\boldsymbol{\theta})q(\boldsymbol{\theta}'|\boldsymbol{\theta}), \pi(\boldsymbol{\theta}')q(\boldsymbol{\theta}|\boldsymbol{\theta}')\right\} = \pi(\boldsymbol{\theta}')P(\boldsymbol{\theta}|\boldsymbol{\theta}'),\end{aligned}$$

where the final equality is by symmetry of the third term above.

6.2.5 A toy example with Metropolis Hastings

To illustrate the basic simplicity of the approach, consider an example for which simulation is certainly not required. Suppose we have 20 independent observations x_i, that can each be modelled as $N(\mu, \sigma^2)$ random variables, and we are interested in inference about μ and σ. In the absence of real prior knowledge about these parameters, suppose that we decide on prior independence and improper prior densities, so that $f(\mu) \propto k$ and $f(\log \sigma) \propto c$ where k and c are constants (values immaterial). We work with $\log \sigma$, because σ is inherently positive.[1]

This specification sets up the Bayesian model. To simulate from the corresponding posterior for the parameters using MH we also need a proposal distribution. In this case let us choose independent scaled t_3 distributions centred on the current parameter values for both parameters. This means that we will propose new parameter values by simply adding a multiple of t_3 random deviates to the current parameter values: this is an example of a *random walk* proposal. The proposed values are then accepted or rejected using the MH mechanism.

Here is some R code to implement this example, using simulated x data from $N(1, 2)$. The parameters are assumed to be in vectors $\boldsymbol{\theta} = (\mu, \log \sigma)^{\mathrm{T}}$ to be stored columnwise in a matrix `theta`.

```
set.seed(1);x <- rnorm(20)*2+1 ## simulated data
n.rep <- 10000; n.accept <- 0
theta <- matrix(0,2,n.rep) ## storage for sim. values
ll0 <- sum(dnorm(x,mean=theta[1,1],
            sd=exp(theta[2,1]),log=TRUE))
for (i in 2:n.rep) { ## The MH loop
  theta[,i] <- theta[,i-1] + rt(2,df=3)*.5 ## proposal
  ll1 <- sum(dnorm(x,mean=theta[1,i],
              sd=exp(theta[2,i]),log=TRUE))
  if (exp(ll1-ll0)>runif(1)) { ## MH accept/reject
    ll0 <- ll1; n.accept <- n.accept + 1 ## accept
  } else theta[,i] <- theta[,i-1] ## reject
}
n.accept/n.rep ## proportion of proposals accepted
```

Working on the log probability scale is a sensible precaution against probabilities underflowing to zero (i.e. being evaluated as zero, merely because they are smaller than the smallest number the computer can represent). The acceptance rate of the chain is monitored, to try to ensure that it is neither too high, nor too low. Too low an acceptance rate is obviously a

[1] Note that a uniform prior on $\log \sigma$ puts a great deal of weight on $\sigma \approx 0$, which can cause problems in cases where the data contain little information on σ.

problem, because the chain then stays in the same state for long periods, resulting in very high correlation and the need for very long runs in order to obtain a sufficiently representative sample. A low acceptance rate may result from a proposal that tries to make very large steps, which are almost always rejected. Less obviously, very high acceptance rates are also a problem because they only tend to occur when the proposal is making very small steps, relative to the scale of variability suggested by the posterior. This again leads to excessively autocorrelated chains and the need for very long runs. It turns out that in many circumstances it is near optimal to accept about a quarter of steps (Roberts et al., 1997). Here we can control the acceptance rate through the standard deviation of the proposal distribution: some experimentation was needed to find that setting this to 0.5 gave an acceptance rate of about 23%.

We need to look at the output. The following code nicely arranges plots of the chain components against iteration, and histograms of the chain components after discarding a burn-in period of 1000 iterations:

```
layout(matrix(c(1,2,1,2,3,4),2,3))
plot(1:n.rep,theta[1,],type="l",xlab="iteration",
                          ylab=expression(mu))
plot(1:n.rep,exp(theta[2,]),type="l",xlab="iteration",
                          ylab=expression(sigma))
hist(theta[1,-(1:1000)],main="",xlab=expression(mu))
hist(exp(theta[2,-(1:1000)]),main="",
                          xlab=expression(sigma))
```

The results are shown in Figure 6.1. The left-hand plots show that the chains appear to have reached a stable state very quickly (rapid convergence) and then move rapidly around that distribution (good mixing). The right-hand histograms illustrate the shape of the marginal distributions of the parameters according to the posterior.

6.2.6 Designing proposal distributions

The catch with Metropolis-Hastings is the proposal distribution. To get the chain to mix well we have to get it right, and for complex models it is seldom the case that we can get away with updating all elements of the parameter vector with independent random steps, all with the same variance, as in the toy example from the last section. In most practical applications, several pilot runs of the MH sampler will be needed to 'tune' the proposal distribution, along with some analysis of model structure. In particular:

1. With simple independent random walk proposals, different standard deviations are likely to be required for different parameters.

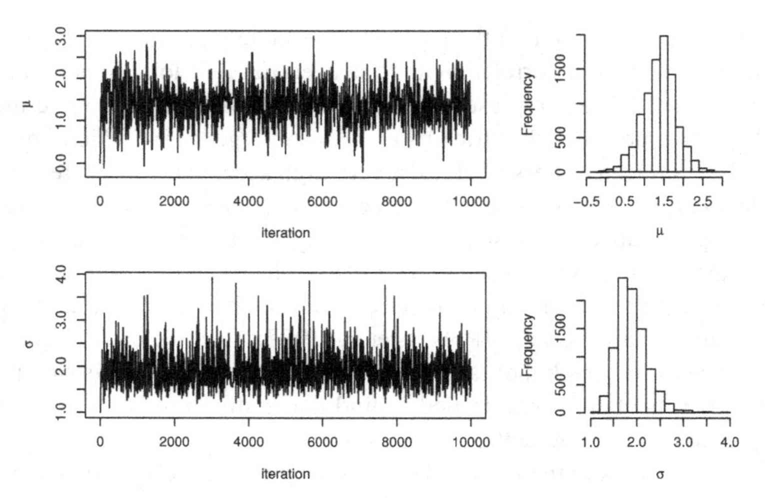

Figure 6.1 Results of the Metropolis-Hastings method applied to the toy model in Section 6.2.5. The left panels show the simulated values for the parameters at each step of the chain, joined by lines. The chains converge very rapidly and mix well in this case. The right panels show histograms of the simulated values of the parameters after discarding the first 1000 steps as burn-in.

2. As its dimension increases it often becomes increasingly difficult to update all elements of $\boldsymbol{\theta}$ simultaneously, unless uselessly tiny steps are proposed. The difficulty is that a purely random step is increasingly unlikely to land in a place where the posterior is non-negligible as dimension increases. In addition it is hard to tune componentwise standard deviations if all elements are proposed together. A solution is to break the proposal down into smaller parts and to only update small mutually exclusive subsets of the parameter vector at each step. The subset to update can be chosen randomly, or we can systematically work through all subsets in some order.[2] This approach only affects the computation of the proposal; the computation of the acceptance ratio is unchanged. But notice that we increase the work required to achieve an update of the whole vector, because the computations required for the accept/reject decision have to be repeated for each subset of parameters.
3. It may be necessary to use correlated proposals, rather than updating each element of $\boldsymbol{\theta}$ independently. Bearing in mind the impractical fact that the perfect proposal would be the posterior itself, it is tempting to

[2] In some rare cases working through all subsets in order can lead to undesirable cycles, or irreversibility of moves: random ordering or random subset selection fixes this.

base the proposal on (6.2), when this is available (or its t_k variant). One can either use it as a static proposal distribution in an MH iteration or simply use a scaled version of the covariance matrix as the basis for taking multivariate normal or t distributed steps (with expectation $\mathbf{0}$). Of course, in many cases we are simulating because other methods fail, and there is little choice but to try and learn the appropriate correlation structure from pilot runs or from the run itself, although this latter option takes us into the realm of adaptive MCMC and beyond the scope of this book.

The issue of proposal design is discussed again in Section 6.5, after examining an example where slow mixing is evident.

6.2.7 Gibbs sampling

When considering the design of proposal distributions, two facts were important. First, it is often necessary to update parameters in blocks. Second, the perfect proposal is the posterior itself: substituting it for q in (6.4), we find that $\alpha = 1$, so that such proposals would always be accepted. On its own the second fact is impractical, but applied blockwise it can result in a very efficient scheme known as Gibbs sampling.[3]

The basic idea is this. Suppose we have a random draw from the joint posterior distribution of $\boldsymbol{\theta}^{[-1]} = (\theta_2, \theta_3, \ldots \theta_q)^{\mathrm{T}}$, and would like a draw from the joint posterior distribution of the whole of $\boldsymbol{\theta}$. This is easy given that $f(\boldsymbol{\theta}|\mathbf{y}) = f(\theta_1|\boldsymbol{\theta}^{[-1]}, \mathbf{y}) f(\boldsymbol{\theta}^{[-1]}|\mathbf{y})$ (see sections 1.4.2 or 1.4.3): simulate θ_1 from $f(\theta_1|\boldsymbol{\theta}^{[-1]}, \mathbf{y})$, append the result to $\boldsymbol{\theta}^{[-1]}$ and we are done. There is nothing special about θ_1 in this process. The same thing would have worked for any other θ_i, or indeed for several θ_i simultaneously. In fact, if we have access to the conditional distributions for all elements of $\boldsymbol{\theta}$ then we could simply cycle through the θ_i updating each in turn and thereby generating a (correlated) sequence of draws from $f(\boldsymbol{\theta}|\mathbf{y})$.

In general then, suppose that the parameter row vector is partitioned into subvectors $\boldsymbol{\theta} = (\boldsymbol{\theta}^{[1]}, \boldsymbol{\theta}^{[2]}, \ldots, \boldsymbol{\theta}^{[K]})$. Further define

$$\tilde{\boldsymbol{\theta}}_j^{[-k]} = (\boldsymbol{\theta}_{j+1}^{[1]}, \boldsymbol{\theta}_{j+1}^{[2]}, \ldots, \boldsymbol{\theta}_{j+1}^{[k-1]}, \boldsymbol{\theta}_j^{[k+1]}, \ldots, \boldsymbol{\theta}_j^{[K]}).$$

Then, given an initial $\boldsymbol{\theta}_1$, J steps of the Gibbs sampler proceed as follows

1. For $j = 1, \ldots, J$ repeat...
2. For $k = 1, \ldots, K$ simulate $\boldsymbol{\theta}_{j+1}^{[k]} \sim f(\boldsymbol{\theta}^{[k]}|\tilde{\boldsymbol{\theta}}_j^{[-k]}, \mathbf{y})$.

[3] In honour of the physical model to which it was first applied.

Notice from the definition of $\tilde{\boldsymbol{\theta}}_j^{[-k]}$ that we always condition on the most recently simulated values.

At this point, the obvious question is how all those conditional distributions are to be found. It is often natural to specify a model in terms of a hierarchy of conditional dependencies, but these dependencies all run in one direction, leaving the problem of working out the conditional dependencies in the other direction. Alternatively, if we attempt to specify the model directly in terms of all its conditional distributions, we will have the no less tricky problem of checking that our specification actually corresponds to a properly defined joint distribution.

Actually, the problem of identifying the conditionals is less daunting than it first seems, and even if we cannot recognise the conditionals as belonging to some standard distribution, it is always possible to devise some way of simulating from them, as the last resort simply using a Metropolis Hastings step for the component. The main trick for recognising conditionals is to use the fact that, for any p.d.f., multiplicative factors that do not involve the argument of the p.d.f. must be part of the normalising constant. To identify a p.d.f. it therefore suffices to recognise its form, to within a normalising constant. The following example helps to clarify this.

6.2.8 Toy Gibbs sampling example

Consider again the toy example from Section 6.2.5, but this time with proper priors on the parameters of the normal model. So we have $n = 20$ observations of a $N(\mu, \phi)$ random variable, where $1/\phi \sim G(a, b)$ (a gamma random variable, with p.d.f. $f(y) = b^a y^{a-1} e^{-by}/\Gamma(a)$) and (independently) $\mu \sim N(c, d)$. a, b, c and d are constants to be specified. The joint density is given by the product of the three densities involved:

$$f(\mathbf{x}, \mu, \phi) \propto \frac{1}{\phi^{n/2}} e^{-\sum_i (x_i - \mu)^2/(2\phi)} e^{-(\mu - c)^2/(2d)} \frac{1}{\phi^{a-1}} e^{-b/\phi}$$

where factors not involving $\mathbf{x}$, ϕ or μ have been omitted because they only contribute to the normalising constant. As we saw in Section 1.4.2, the conditional densities are proportional to the joint density, at the conditioning values. So we can read off the conditional for $1/\phi$, again ignoring factors that do not contain ϕ (and hence contribute only to the normalising constant):

$$f(1/\phi|\mathbf{x}, \mu) \propto \frac{1}{\phi^{n/2+a-1}} e^{-\sum_i (x_i - \mu)^2/(2\phi) - b/\phi}.$$

If this is to be a p.d.f., then it is recognisable as a $G(n/2+a-1, \sum_i(x_i - \mu)^2/2 + b)$ p.d.f.

The conditional for μ is more tedious,

$$
\begin{aligned}
f(\mu|\mathbf{x}, \phi) &\propto e^{-\sum_i(x_i-\mu)^2/(2\phi)-(\mu-c)^2/(2d)} \\
&\propto e^{-(n\mu^2-2\bar{x}n\mu)/(2\phi)-(\mu^2-2\mu c)/(2d)} \\
&= e^{-\frac{1}{2\phi d}(dn\mu^2-2\bar{x}dn\mu+\phi\mu^2-2\mu\phi c)} = e^{-\frac{dn+\phi}{2\phi d}\left(\mu^2-2\mu\frac{dn\bar{x}+\phi c}{dn+\phi}\right)} \\
&\propto e^{-\frac{dn+\phi}{2\phi d}\left(\mu-\frac{dn\bar{x}+\phi c}{dn+\phi}\right)^2},
\end{aligned}
$$

where terms involving only $\sum_i x_i^2$ and c^2 were absorbed into the normalising constant at the second '$\propto$', and the constant required to complete the square at the final '$\propto$' has been taken from the normalising constant. From the final line, we see that

$$\mu|\mathbf{x}, \phi \sim N\left(\frac{dn\bar{x} + \phi c}{dn + \phi}, \frac{\phi d}{dn + \phi}\right).$$

Now it is easy to code up a Gibbs sampler:

```
n <- 20;set.seed(1);x <- rnorm(n)*2+1 ## simulated data

n.rep <- 10000;
thetag <- matrix(0,2,n.rep)

a <- 1; b <- .1; c <- 0; d <- 100 ## prior constants
xbar <- mean(x)               ## store mean
thetag[,1] <- c(mu <- 0,phi <- 1) ## initial guesses
for (j in 2:n.rep) { ## the Gibbs sampling loop
  mu <- rnorm(1,mean=(d*n*xbar+phi*c)/(d*n+phi),
                sd=sqrt(phi*d/(d*n+phi)))
  phi <- 1/rgamma(1,n/2+a-1,sum((x-mu)^2)/2+b)
  thetag[,j] <- c(mu,phi) ## store results
}
```

The equivalent of Figure 6.1 is shown in Figure 6.2. Notice the rather limited effect of the change in prior between the two figures. This is because even the proper priors used for the Gibbs sampler are very vague, providing very limited information on the probable parameter values (plot the $\Gamma(1, .1)$ density to see this), while the data are informative.

6.2.9 Metropolis within Gibbs example

As mentioned previously, we can substitute MH steps for any conditional that we cannot obtain, or cannot be bothered to obtain. Recycling the toy

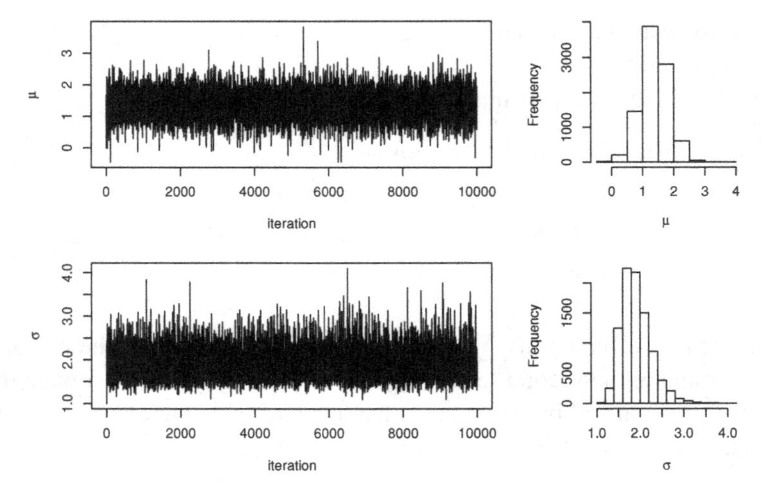

Figure 6.2 Results of Gibbs sampling applied to the toy model in Section 6.2.8. The left panels show the simulated values for the parameters at each step of the chain, joined by lines. The right panels show histograms of the simulated values of the parameters after discarding the first 1000 steps.

example one more time, let us suppose that the conditional for μ is simply too much effort:

```
a <- 1; b <- .1; c <- 0; d <- 100
mu <- 0; phi <- 1; n.accept <- 0
thetamg[,1] <- c(mu,phi)
for (j in 2:n.rep) {
  mup <- mu + rnorm(1)*.8 ## proposal for mu
  log.a <- sum(dnorm(x,mup,sqrt(phi),log=TRUE)) +
          dnorm(mup,c,sqrt(d),log=TRUE) -
          sum(dnorm(x,mu,sqrt(phi),log=TRUE)) -
          dnorm(mu,c,sqrt(d),log=TRUE)
  if (runif(1) < exp(log.a)) { ## MH accept?
    mu <- mup;n.accept <- n.accept + 1
  }
  ## Gibbs update of phi...
  phi <- 1/rgamma(1,n/2+a-1,sum((x-mu)^2)/2+b)
  thetamg[,j] <- c(mu,phi) ## store results
}
n.accept/n.rep
```

The acceptance rate is about 50% (actually about optimal in the single-parameter case). Figure 6.3 shows the results. The μ chain is not quite as impressive as in the pure Gibbs case, but beats the pure MH results.

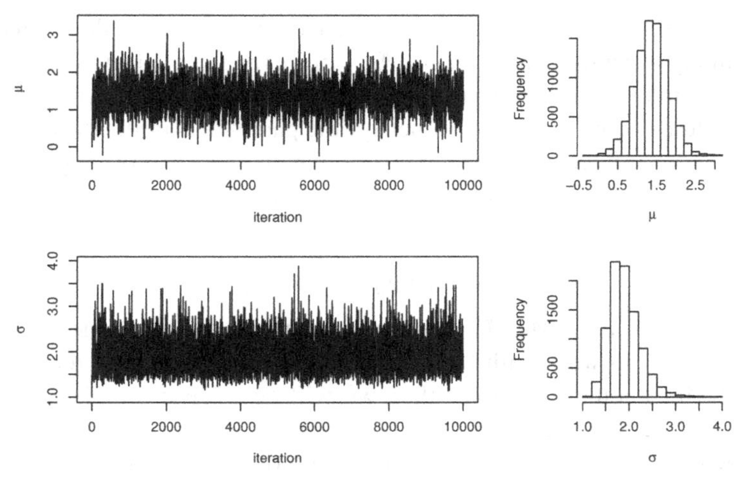

Figure 6.3 Results of Metropolis within Gibbs sampling applied to the toy model of Section 6.2.8. The left panels show the simulated values for the parameters at each step of the chain, joined by lines. The right panels show histograms of the simulated values of the parameters after discarding the first 1000 steps as a burn-in period.

6.2.10 Limitations of Gibbs sampling

Gibbs sampling largely eliminates the difficulty of choosing a good proposal that complicates Metropolis Hastings, but this is not quite the free lunch that it might appear. The catch is that Gibbs sampling produces slowly moving chains if parameters have high posterior correlation, because sampling from the conditionals then produces very small steps. Sometimes updating parameters in blocks or re-parameterising to reduce posterior dependence can then help to improve mixing. The other practical consideration is that if improper priors are used with Gibbs sampling then it is important to check that the posterior is actually proper: it is not always possible to detect impropriety from the output of the sampler.

6.2.11 Random effects

A beauty of the Bayesian simulation approach is that there is almost nothing to say about random effects: for simulation purposes they are simply treated as if they were parameters. The point here is that if we have a sample from the joint posterior $f(\boldsymbol{\theta}, \mathbf{b}|\mathbf{y})$ of the parameters and random effects, then simply discarding the random effects from the sample leaves us with a sample from the marginal posterior density $f(\boldsymbol{\theta}|\mathbf{y})$. The only caveat is that one would not usually specify values for the parameters of the distribution of the random effects directly, but instead would choose to place priors on those parameters (often of a rather vague nature).

6.2.12 Checking for convergence

The great appeal of MCMC methods is their impressive generality. In principle we can work with almost any model, and if we are prepared to simulate for enough iterations, then we will generate a sample from its posterior. The difficulty is in identifying what is enough iterations. Well-designed samplers for relatively benign posteriors may require only several thousand or several hundred thousand iterations. In other situations all the computer power on earth running for the age of the universe might be required to adequately sample from the posterior: for example, when the posterior consists of several well-separated and compact modes in a high-dimensional space, where proposing a move that will take us from one mode to another is all but impossible, let alone doing it often enough to sample from the modes in the correct proportion. To appreciate the problem, suppose that your posterior is like the final plot in Section 5.6.1, but without the middle peak. In the MH setting, any proposal with a nonvanishing chance of making the transition from one mode to another would lead to tiny acceptance rates. In the Gibbs setting the conditionals would look like the penultimate plot in Section 5.6.1, and it would be virtually impossible for Gibbs sampling to make the transition from one peak to another.

So it is important to check for apparent convergence of MCMC chains. Obvious checks are the sort of plots produced in the left-hand panels of Figures 6.1 to 6.3, which give us some visual indication of convergence and how well the chain is mixing. If there is any suspicion that the posterior could be multimodal, then it is sensible to run multiple chains from radically different starting points to check that they appear to be converging to the same distribution. If interest is actually in some scalar valued function of the parameters $h(\boldsymbol{\theta})$, then it makes sense to produce the plots and other diagnostics directly for this quantity.

One step up in sophistication from the simple trace plots is to examine how specific quantiles of the sample, up to iteration j, behave when plotted against j. For example, the following code produces plots that overlay the

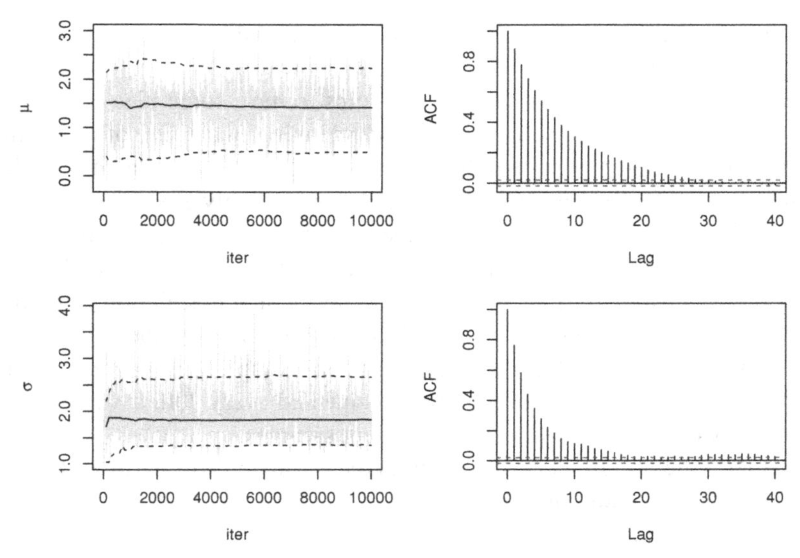

Figure 6.4 Basic MCMC checking plots, relating to the example from Section 6.2.5 as discussed in Section 6.2.12. The left panels show trace plots for the two chain components (grey) overlaid with the evolution of the chain median (continuous), 0.025 and 0.975 quantiles (dashed). Everything is stabilising nicely here. The right-hand panels show the equivalent autocorrelation function plots for the chains, illustrating the substantial degree of autocorrelation when using this sampler, which is higher for the μ component than for the σ component.

0.025, 0.5 and 0.975 quantiles of the sample so far, on top of a simple line plot of the chain's progress:

```
qtplot <- function(theta,n.plot=100,ylab="") {
## simple MCMC chain diagnostic plot
  cuq <- Vectorize(function(n,x) ## cumul. quantile func.
       as.numeric(quantile(x[1:n],c(.025,.5,.975))),
       vectorize.args="n")
  n.rep <- length(theta)
  plot(1:n.rep,theta,col="lightgrey",xlab="iter",
                                    ylab=ylab,type="l")
  iter <- round(seq(1,n.rep,length=n.plot+1)[-1])
  tq <- cuq(iter,theta)
  lines(iter,tq[2,])
  lines(iter,tq[1,],lty=2);lines(iter,tq[3,],lty=2)
}
```

A call to `qtplot(theta[1,],ylab=expression(mu))` produces the upper left plot in Figure 6.4, with a slightly modified call producing the lower left plot. In both cases it appears that the median and both the other quantiles stabilise rapidly.

For slightly more formal checking it helps to have some idea of the *effective sample size* of the chains. Roughly speaking, what size of independent sample from $f(\boldsymbol{\theta}|\mathbf{y})$ would be equivalent to the correlated sample from our MCMC scheme? Or, if we were to retain only every k^{th} sample from the chain, how large would k have to be before we could reasonably treat the resulting thinned sample as approximately independent? To answer these questions it is helpful to examine the autocorrelation function (ACF) of the chain components, as shown in the plots on the right hand side of Figure 6.4, which are produced by, for example, `acf(theta[1,])` in R. Apparently retaining every 25th μ value and every 20th σ value from the chain would give us almost independent samples.

Actually the `acf` function also returns the estimated correlations at each lag (silently). For example,

```
> mu.ac <- acf(theta[1,])[[1]][,,1];mu.ac
 [1] 1.0000000 0.8831924 0.7777484 0.6863556 0.6096274
 [6] 0.5406225 0.4834136 0.4303171 0.3796966 0.3389512
  .    .         .         .         .         .
```

The *autocorrelation length* associated with a chain is defined as twice the sum of the correlations minus 1. The summation is strictly over all lags up to infinity, but in practice we can sum over the lags up to the point at which autocorrelation appears to have vanished (see R package `coda` for a better method). The corresponding effective sample size is then defined as the sequence length divided by the autocorrelation length. For example,

```
> acl <- 2*sum(mu.ac)-1; acl
[1] 16.39729
> n.rep/acl ## effective sample size
[1] 609.8569
```

So the effective sample size for the μ component is about 600 (although usually we would discard the burn-in period before computing this). Repeating the exercise for σ gives an autocorrelation length of around 10 and an effective sample size of about 1000. For the Metropolis within Gibbs sampler, considered previously, the autocorrelation length is only 6 for μ and 1.3 for σ. For pure Gibbs the figure for μ drops to close to 1.

Armed with this information, more formal tests are possible. For example, given multiple chains, we can subsample to obtain approximately independent samples between each chain and then apply ANOVA methods to see if there appear to be variance components associated with the difference between chains. With a single chain, we might want to divide the apparently converged chain into two parts and formally test whether the two samples appear to come from the same distribution. A two-sample Kolmogorov-Smirnov test is appropriate here, provided that the two samples are of independent draws from the two distributions, so again subsampling is needed. Here is an example, for the simple μ chain:

```
> th0 <- theta[1,1001:550]
> th1 <- theta[1,5501:10000]
> ind <- seq(1,4500,by=16) ## subsampling index
> ks.test(th0[ind],th1[ind])

        Two-sample Kolmogorov-Smirnov test

data:  th0[ind] and th1[ind]
D = 0.0745, p-value = 0.4148
alternative hypothesis: two-sided
```

With a p-value of 0.4 there is no evidence for a difference in distribution between the two halves of the chain, so this result provides no reason to doubt convergence. The exact choice of sampling interval is not important: increasing the sampling interval to 25, as might be implied by simply examining the ACF, leads to the same conclusion. However, using a much lower sampling rate is a disaster: failure to subsample at all completely violates the independence assumption of the test, resulting in a computed p-value around 10^{-13}, and even a sampling interval of five results in an erroneously low p-value of 0.016.

This section only scratches the surface of convergence checking. See Robert and Casella (2009, Ch. 8) for more information, and the `coda` package in R for an extensive set of checking functions (Plummer et al., 2006).

6.3 Interval estimation and model comparison

Given reliable posterior simulations from a chain, interval estimates and quantities for model comparison can be computed. The former is straightforward, because intervals can be based directly on the observed quantiles of the simulated parameters. For example, with the simple toy model of Section 6.2.9, it is easy to produce 95% *credible intervals* (CIs) for μ and σ, as follows (discarding the first 1000 samples as burn-in):

```
quantile(thetamg[1,-(1:1000)],c(0.025,0.975)) ## CI mu
quantile(thetamg[2,-(1:1000)]^.5,c(0.025,0.975)) # CI sig
```

which yields $0.52 < \mu < 2.22$ and $1.39 < \sigma < 2.67$. The next subsection discusses model comparison.

6.3.1 Computing the marginal likelihood and DIC

As discussed in Section 2.5.2 Bayesian model comparison presents some fundamental difficulties, which it is important to be aware of before computing anything. For now, suppose that our models are specified with meaningful priors, so that marginal likelihood can form a meaningful basis for model comparison. In that case importance sampling, as introduced in Section 6.1, provides a reasonable way to compute the marginal likelihood, $f(\mathbf{y}) = \int f(\mathbf{y}|\boldsymbol{\theta})f(\boldsymbol{\theta})d\boldsymbol{\theta}$, which is the basis for computing the Bayes factor (introduced in Section 2.5.2) for model comparison. Recall that the idea of importance sampling is to generate n random vectors $\boldsymbol{\theta}_i$ from some suitable density $\propto g(\boldsymbol{\theta})$, and then use the estimate

$$\hat{f}(\mathbf{y}) = \frac{\sum_i f(\mathbf{y}|\boldsymbol{\theta}_i)f(\boldsymbol{\theta}_i)/g(\boldsymbol{\theta}_i)}{\sum_i f(\boldsymbol{\theta}_i)/g(\boldsymbol{\theta}_i)},$$

where the denominator is simply replaced by n if the density g is properly normalised. It is tempting to use the results of MCMC sampling directly here, and set $g(\boldsymbol{\theta}_i) = f(\mathbf{y}|\boldsymbol{\theta}_i)f(\boldsymbol{\theta}_i)$, in which case

$$\hat{f}(\mathbf{y}) = \frac{n}{\sum_i 1/f(\mathbf{y}|\boldsymbol{\theta}_i)},$$

which is the harmonic mean of the likelihood over the simulations. Unfortunately this simple estimate is of poor quality. It need not have finite variance, and its realised behaviour is often bizarre.[4] The problem is that the harmonic mean is dominated by the smallest values of $f(\mathbf{y}|\boldsymbol{\theta}_i)$ in the sample, and the greater the sample, the smaller the smallest values become. In consequence this estimate can depend strongly and systematically on the simulation length n. This problem would not occur if we simply set $g(\boldsymbol{\theta}_i) = f(\boldsymbol{\theta}_i)$, but in the case of a highly informative likelihood and/or diffuse priors such a scheme would put most of the simulated $\boldsymbol{\theta}_i$ where the integrand is negligible, resulting in high estimator variance.

An obvious solution is to base sampling on a mixture of the two approaches; that is, we simulate a sample from the prior and a sample from the posterior and treat the combined sample as coming from the mixture distribution $g(\boldsymbol{\theta}) = \alpha f(\boldsymbol{\theta}|\mathbf{y}) + (1-\alpha)f(\boldsymbol{\theta})$, where $0 < \alpha < 1$ and g is now properly normalised. The difficulty, of course, is that $f(\boldsymbol{\theta}|\mathbf{y})$ involves

[4] The poor performance is initially confusing, since from basic importance sampling theory the proposal distribution appears to be ideal, but the pathology creeps in because of the need to normalise the importance weights in this case.

the very normalising constant that we are trying to find, but plugging the estimate of $\hat{f}(\mathbf{y})$ into the importance sampling estimate yields

$$\hat{f}(\mathbf{y}) = \frac{1}{n}\sum_{i=1}^{n} \frac{f(\mathbf{y}|\boldsymbol{\theta}_i)f(\boldsymbol{\theta}_i)}{\alpha f(\mathbf{y}|\boldsymbol{\theta}_i)f(\boldsymbol{\theta}_i)/\hat{f}(\mathbf{y}) + (1-\alpha)f(\boldsymbol{\theta}_i)}, \tag{6.5}$$

which can be solved, numerically, for $\hat{f}(\mathbf{y})$.[5]

To use this importance sampling based estimate in practice requires that some care is taken to avoid underflow or overflow problems. Letting $c = \log \hat{f}(\mathbf{y})$ and $a_i = \log f(\mathbf{y}|\boldsymbol{\theta}_i)$, (6.5) can be rewritten as

$$\log \sum_{i}^{n} \left\{\alpha e^{-\beta} + (1-\alpha)e^{c-a_i-\beta})\right\}^{-1} - \beta - \log n = 0,$$

where β is an arbitrary constant, which can be set to a nonzero value, if necessary, to avoid overflow problems. Note that in practice the root c of (6.5) may be unbounded below computationally, if the posterior probability of the draws from the prior is vanishingly small computationally. This can occur in high-dimensional settings and when using vague or uninformative priors, but in the latter case the marginal likelihood and Bayes factors should anyway not be computed for the reasons given in Section 2.5.2.

To see this in action, suppose that we have generated 20,000 samples using the Gibbs sampling code in Section 6.2.8, and that these samples are stored in the two row matrix `thetag`. Now add 20,000 samples from the prior, and compute the log likelihood for each sample:

```
n.prior <- 20000
thetap <- matrix(0,2,n.prior)
thetap[1,] <- rnorm(n.prior,c,sqrt(d))
thetap[2,] <- rgamma(n.prior,a,b)
th <- cbind(thetag,thetap) ## combined sample
alpha=ncol(thetag)/ncol(th)
lfy.th <- colSums(matrix(dnorm(x,rep(th[1,],each=n),
     rep(sqrt(th[2,]),each=n),log=TRUE),n,n.rep+n.prior))
```

With these ingredients we can now solve for $c = \log \hat{f}(\mathbf{y})$. The following function implements the stabilised version of (6.5), computing a value for

[5] It is easy to prove that the equation always has a single finite root by defining $k = 1/\hat{f}(\mathbf{y})$ and then considering where the curve of $1/k$ versus k cuts the (monotonic) curve of the right hand side against k. That the root is the marginal likelihood in the large sample limit follows from the unbiasedness and consistency of importance sampling when the true $f(\mathbf{y})$ is substituted in the right hand side.

β that should reduce underflow or overflow, if this is necessary:

```
fyf <- function(lfy,lfy.th,alpha,big=280) {
  ## log f(y) - log f(y|theta_i) = c - a_i ...
  ac <- lfy - lfy.th
  if (min(ac) < -big||max(ac) > big) { ## overflow?
    beta <- sum(range(ac))/2
    if (beta > big) beta <- big
    if (beta < -big) beta <- -big
  } else beta <- 0
  n <- length(lfy.th)
  ac <- ac - beta
  ind <- ac < big ## index non-overflowing ac values
  log(sum(1/(alpha*exp(-beta) + (1-alpha)*exp(ac[ind]))))-
       beta - log(n)
}
```

The `uniroot` function in R can be used to solve for the log of $\hat{f}(\mathbf{y})$, as follows:

```
>uniroot(fyf,interval=c(-100,0),lfy.th=lfy.th,alpha=alpha)
$root
[1] -44.64441
```

So the log of the marginal likelihood is approximately -44.6 here.[6] If we had two models to compare, then we could compute the log marginal likelihood for the second model in the same way, form the log Bayes factor, and then refer to Section 2.5.2 for interpretation of the result. Note, however, that this example is really only useful for illustrating the computations: the vague priors in this model cannot be treated as the sort of meaningful prior information that would justify use of the Bayes factor for serious model comparison.

Computations for the fractional Bayes factor

Recall from (2.7) in Section 2.5.2 that to compute the fractional Bayes factor requires that the (estimated) marginal likelihood be divided by (an estimate of) $\int f(\mathbf{y}|\boldsymbol{\theta})^b f(\boldsymbol{\theta})d\boldsymbol{\theta}$ where b is constant in $(0, 1)$. Let us call the resulting quantity the 'fractional marginal likelihood' here. Given the preceding computations it is easy to reuse (6.5) and estimate the required

[6] This example can also be used to illustrate the problem with the harmonic mean estimator based on the chain samples alone: running the chain for 10 times as many iterations increases the harmonic mean estimate of the marginal likelihood by a factor of around 10.

integral using

$$\int f(\mathbf{y}|\boldsymbol{\theta})^b f(\boldsymbol{\theta}) d\boldsymbol{\theta} \simeq \frac{1}{n} \sum_{i=1}^{n} \frac{f(\mathbf{y}|\boldsymbol{\theta}_i)^b}{\alpha f(\mathbf{y}|\boldsymbol{\theta}_i)/\hat{f}(\mathbf{y}) + (1-\alpha)}$$

Setting $b = 0.3$, here is some R code for the computation:

```
lfy <- uniroot(fyf,interval=c(-100,0),lfy.th=lfy.th,
                  alpha=alpha)$root
lfyb <- log(mean(exp(.3 * lfy.th -
                  log(.5 * exp(lfy.th-lfy) + .5))))
frac.ml <- lfy - lfyb ## log 'fractional ML'
```

The result is -29.3. To illustrate the robustness of the fractional approach, the prior parameters in this example can be changed to `b=0.01` and `d=1000` (making the priors more vague). The marginal likelihood then drops from -44.6 to -47.4, whereas the fractional version only drops to -29.4.

A crude Laplace approximate marginal likelihood estimate

In the large sample limit with informative data, the posterior covariance matrix is the inverse Hessian of the log likelihood, which dominates the log prior. Hence if $\hat{\boldsymbol{\Sigma}}$ is the estimated covariance matrix of $\boldsymbol{\theta}$ from the chain, while $\hat{f}$ is the largest value of $f(\mathbf{y}|\boldsymbol{\theta})f(\boldsymbol{\theta})$ observed in the chain, then applying a Laplace approximation as in Section 5.3.1 yields the rough approximation

$$\log \hat{f}(\mathbf{y}) \simeq \log \hat{f} + p \log(2\pi)/2 + \log|\hat{\boldsymbol{\Sigma}}|/2,$$

where $p = \dim(\theta)$. This approximation is computable directly from a sample from the posterior. For example, continuing the example from the previous subsection,

```
> V <- cov(t(thetag))
> lfyth <- lfy.th[1:n.rep] +
+             dnorm(thetag[1,],c,sqrt(d),log=TRUE) +
+             dgamma(thetag[2,],a,b,log=TRUE)
> max(lfyth) + log(2*pi) + sum(log(diag(chol(V))))
[1] -44.44495
```

which is comparable with the previous estimate. This approach only works in circumstances in which a Laplace approximation is expected to work, so it will only be useful when the posterior has a single important mode, which can be reasonably approximated by a Gaussian.

Neither of the simple methods presented here is likely to work well in very complex modelling situations. In such cases (and again assuming that meaningful priors have been used) more sophisticated methods will be needed: Friel and Pettitt (2008) is a good place to start.

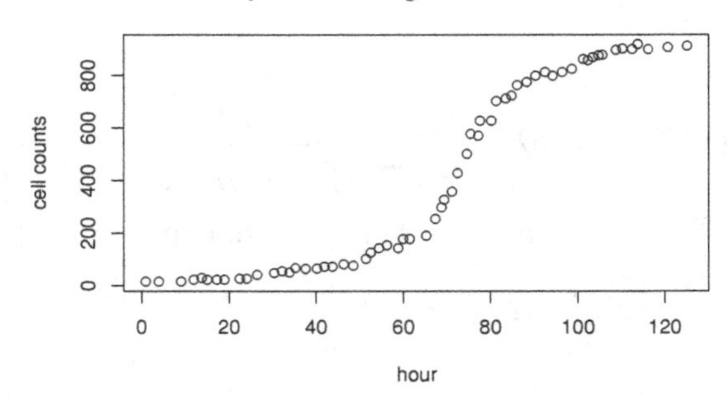

Figure 6.5 Algal cell counts in samples taken from a laboratory chemostat experiment, against hour.

Computing the DIC

Relative to the marginal likelihood, computing the DIC is very easy, and it can legitimately be used with vague priors. Continuing the same example,

```
> Dthbar <- -2*sum(dnorm(x,mean(thetag[1,]),
+                  mean(thetag[2,])^.5,log=TRUE))
> pD <- mean(-2*lfy.th[1:n.rep]) - lfy.thbar
> DIC <- Dthbar + 2*pD; DIC; pD
[1] 83.79534
[1] 1.851937
```

So the DIC is 83.8 and the effective degrees of freedom, p_D, is 1.85. As with AIC we would favour models with a smaller DIC.

6.4 An MCMC example: algal growth

This section covers a non trivial example of using Metropolis Hastings sampling. Figure 6.5 shows counts of algal cells in samples drawn from a laboratory chemostat experiment. A possible model for the population growth in the chemostat is that it follows a self-damping growth model, such as

$$N_{t+1} = e^r N_t e^{-N_t/K+e_t}, \quad e_t \sim N(0, \sigma_e^2), \tag{6.6}$$

where the independent e_t terms reflect the fact that the population growth will not be fully deterministic. This model would usually operate with a fixed timestep (e.g. t might index hour, or two hour period). If we want to estimate N between N_t and N_{t+1} then we might use linear interpolation. The cell population, y, is then modelled as being a noisy observation of the underlying population N, perhaps Gaussian, with unknown variance, σ^2.

The first thing to note is that the data plotted in Figure 6.5 are not evenly spaced in time. The spacing ranges from 0.4 to 5.1 hours. Hence we will need to interpolate the solution to (6.6) in order to use this model for the data. Because MCMC sampling will involve doing essentially the same interpolation repeatedly, it makes sense to try and do it efficiently. So here is a function that supposes that you have evaluated N at m discrete times, starting at `t0` and spaced `dt` apart, and that you want to use linear interpolation to estimate the value of N at the times given in `t`. The following routine returns vectors `im`, `ip`, `wm` and `wp`, each of the same length as `t`, such that if `N` is the m vector of evenly spaced N values, `N[im] *wm + N[ip] *wp` gives the vector of interpolated N estimates, corresponding to `t`. It also returns `dt` and an appropriate value for m:

```
lint <- function(t,t0=0,dt=1) {
## produce interpolation indices and weights.
  n <- length(t)
  ts <- seq(t0,max(t),by=dt)
  ts <- c(ts,max(ts)+dt)
  m <- length(ts)
  im <- floor((t-t0)/dt)+1;ip <- im+1;ip[ip>m] <- m
  list(im=im,ip=ip,wm=(ts[ip] - t)/dt,
          wp=(t - ts[im])/dt,m=m,dt=dt)
}
```

Armed with this function it is now possible to write a function to evaluate the joint density of the cell count data, random effects and model parameters. The obvious way to approach this is to write down the joint density of $\mathbf{y}$ and $\mathbf{e}$, but actually this can make sampling very difficult indeed. Because early e_t values affect the whole subsequent sequence of N_t values, it can be very difficult to propose acceptable moves. In fact for high r values it is essentially impossible. However, there is no such problem if we work directly in terms of the log state, $n_t = \log N_t$. It is then easy to establish a one-to-one transformation between the state vector $\mathbf{n}$ and the random effect vector $\mathbf{e}$, (with determinant 1) and hence to evaluate the joint density of the data, state vector and parameters. This leads to a function like the following (where improper uniform priors are assumed on all log parameters):

```
lfey <- function(theta,n,y,li) {
## function evaluating log p.d.f. of y, n and
## theta of Ricker model
  theta <- exp(theta) ## parameters are intrinsically +ve
  r <- theta[1]; n0 <- theta[2]; K <- theta[3];
  sigma.e <- theta[4]; sigma <- theta[5]
  n.n <- length(n); ind <- 1:(n.n-1);
  ## state to r.e. transform...
  e <- c(n[1]-log(n0),n[ind+1]-n[ind]-r+exp(n[ind])/K)
  f.ne <- sum(dnorm(e,0,sigma.e,log=TRUE)) ## r.e. density
  mu <- exp(li$wm*n[li$im] + li$wp*n[li$ip]) # interpolate
  f.y <- sum(dnorm(y,mu,sigma,log=TRUE)) ## f(y|n)
  f.y + f.ne ## joint log density
}
```

Finding a proposal to update the parameters and whole state vector in one go requires further work, as described in Sections 6.5.3 and 6.5.4. A simpler approach is to make proposals for each element of the parameter and state vector separately, either choosing the element to update at random in each step or working through each element in turn at every step. The latter approach is the basis for the following code. It is assumed that the data are in a data frame called `alg`.

```
li <- lint(alg$hour,t0=0,dt=4) ## interpolation weights
## Intial values...
n0 <- 10;r <- .3; K <- 3000; sig.b <- .2; sigma <- 10
theta <- log(c(r,n0,K,sig.b,sigma)) ## parameter vector

## get initial state by interpolating data...
n <- log(c(alg$cell.pop[1],approx(alg$hour,alg$cell.pop,
          1:(li$m-2)*li$dt)$y,max(alg$cell.pop)))

n.mc <- 150000 ## chain length
th <- matrix(0,length(theta),n.mc)
y <- alg$cell.pop

a.th <- rep(0,length(theta)); a.n <- 0 ## accept counter
sd.theta <- c(.2,.5,.3,.3,.2); sd.n <- .05 ## prop. sd
ll <- c(-Inf,-Inf,-Inf,log(.03),log(5)) ## low param lim
ul <- c(Inf,Inf,log(25000),Inf,Inf) ## upper param lim

lf0 <- lfey(theta,n,y,li)
for (i in 1:n.mc) { ## mcmc loop
  for (j in 1:5) { ## update parameters
    theta0 <- theta[j]
    theta[j] <- theta[j] + rnorm(1)*sd.theta[j]
    lf1 <- lfey(theta,n,y,li)
    if (runif(1)<exp(lf1-lf0)&&ll[j]<theta[j]
                            &&ul[j]>theta[j]) { ## accept
      lf0 <- lf1
      a.th[j] <- a.th[j] + 1
    } else { ## reject
      theta[j] <- theta0
      lf1 <- lf0
    }
  } ## parameters updated
  for (j in 1:li$m) { ## update state
    nj <- n[j]
    n[j] <- n[j] + rnorm(1)*sd.n
    lf1 <- lfey(theta,n,y,li)
    if (runif(1)<exp(lf1-lf0)) { ## accept
      lf0 <- lf1
      a.n <- a.n + 1
    } else { ## reject
      n[j] <- nj
```

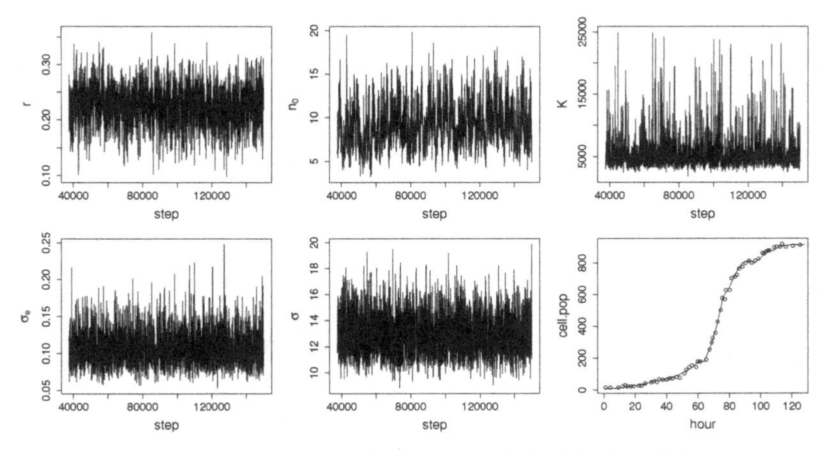

Figure 6.6 First five panels from the top left: Algal model MCMC chains for the model of Section 6.4 sampled every 25 steps. Final panel, bottom right: the state at the final step of the simulation (black line) overlaid on the cell data (open circles).

```
      lf1 <- lf0
    }
  } ## states updated
  th[,i] <- theta ## store theta
  if (i%%1000==0) cat(".")
} ## end of mcmc loop
```

Notice the `ll` and `ul` vectors, which serve to impose lower and upper bounds, respectively, on some parameters (if both are present, then the prior becomes a proper uniform p.d.f.). Also `a.n` and `a.th` are used to monitor acceptance rates, which allowed `sd.theta` and `sd.n` to be tuned in pilot runs to achieve acceptance rates around the 50% level appropriate for single-element updates.

Figure 6.6 shows the results of the simulation, along with the simulated population vector `n` at the final step, overlaid on the raw data. The initial population n_0 does not appear to be well identified, but otherwise mixing seems reasonable. Using `effectiveSize(mcmc(th[i,ind]))` from the `coda` package, the effective sample size for r is around 1800, for n_0 it is around 370, and for the other parameters it is more than 3000. `n` can be transformed into a vector of e_t values, and we can compute residuals to check the sampling error distribution. Here is some code to do this:

```
n0 <- exp(theta[2]); r <- exp(theta[1])
K <- exp(theta[3]); n.n <- length(n)
ind <- 1:(n.n-1);
```

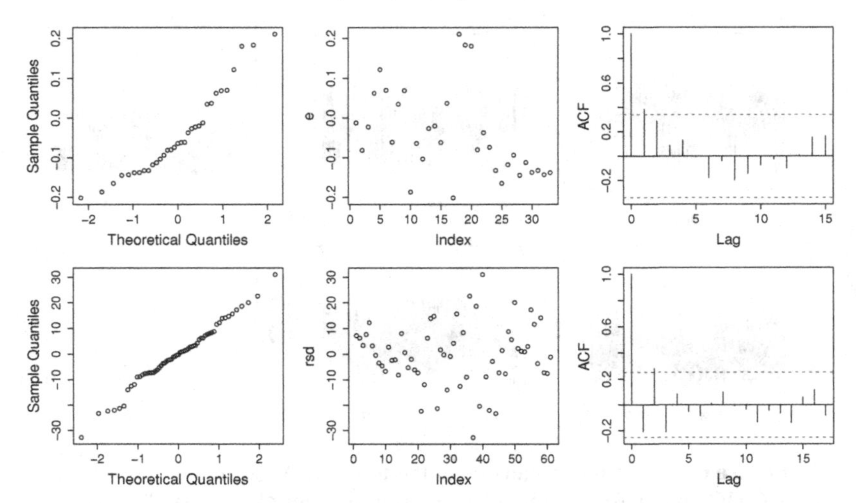

Figure 6.7 Normal QQ-plots, residuals against order, and ACF of residuals for e_t (top row) and residuals (bottom row). The middle plot on the bottom row indicates that the measurement error model is not quite right at the start of the experiment: the error variance is not constant. The top row also indicates slightly heavier than normal tails in the e_t.

```
e <- c(n[1]-log(n0),n[ind+1]-n[ind]-r+exp(n[ind])/K)
rsd <- y - exp(n[li$ip]*li$wp+n[li$im]*li$wm)
par(mfrow=c(2,3),mar=c(5,5,1,1))
qqnorm(e); plot(e); acf(e)
qqnorm(rsd); plot(rsd); acf(rsd)
```

Figure 6.7 shows the results. Clearly, the measurement error model is not quite right, but otherwise the model assumptions seem reasonably plausible. Finally, here is a 90% credible interval for r (having discarded the first 30000 simulations as burn-in):

```
> exp(quantile(th[1,30000:n.mc],c(.05,.95)))
       5%       95%
0.1689796 0.2817391
```

6.5 Geometry of sampling and construction of better proposals

The algal growth example of the previous section highlights the difficulty of constructing good proposals. To be able to tune the proposals and get reasonable movement, it was necessary to resort to single-component updates. This increased the cost of each complete update but still gave slow

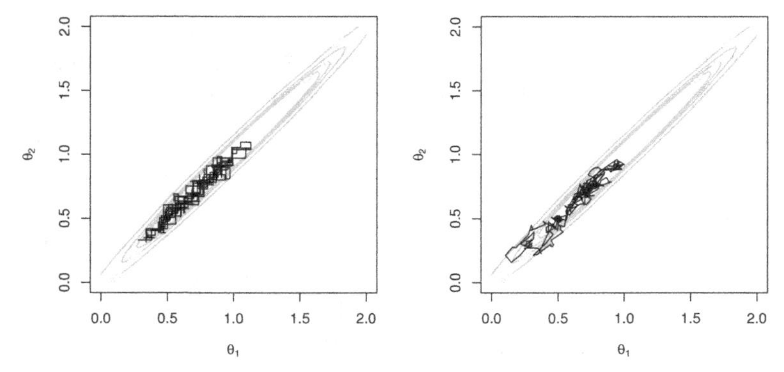

Figure 6.8 Effect of posterior correlation on MCMC mixing. In both panels the posterior for $\boldsymbol{\theta}$ is contoured in grey: over most of the area shown, the posterior density is close to zero. The left panel shows the progress of the chain from the point $(0.5, 0.5)$, when θ_1 and θ_2 are updated separately. The right panel shows the progress when $\boldsymbol{\theta}$ is updated jointly. Both chains were run for 400 steps, and the proposal standard deviations were tuned to obtain optimal acceptance rates. The chains make slow progress because steps have to be kept small in order to keep the $\boldsymbol{\theta}$ within the area of high posterior probability.

mixing. To design more efficient proposals it is necessary to understand how the twin curses of dimensionality and correlation affect proposals.

6.5.1 Posterior correlation

Suppose that the posterior density of $\boldsymbol{\theta}$ (including any random effects) implies that the elements of $\boldsymbol{\theta}$ are highly non-independent. In this case both single-component updates and joint updates, based on independent jumps for each component, will give slow mixing. Neither is able to take big steps without frequently proposing highly improbable moves. The issue is illustrated in Figure 6.8. Neither componentwise nor joint but independent, proposals use the correlation structure in the posterior, with the result that they can only take small steps if the region of negligible posterior probability is to be avoided. As a result the chain moves slowly.

The issue with correlation is a fundamental limitation for Gibbs sampling (although re-parameterisation and/or updating parameters in blocks can often help), but for Metroplois-Hastings sampling it is often possible to exploit the correlation structure in the posterior to improve the proposals. The simplest approaches approximate the posterior density by a multivariate normal density, which is then used as the basis for making proposals.

Two alternatives for obtaining a multivariate normal approximation are to use (6.2) from Section 6.1, or to run a pilot chain, from which to extract $\hat{\boldsymbol{\mu}} \simeq E(\boldsymbol{\theta})$ and $\hat{\boldsymbol{\Sigma}} \simeq \text{cov}(\boldsymbol{\theta})$, so that the approximation is

$$\boldsymbol{\theta}|\mathbf{y} \sim N(\hat{\boldsymbol{\mu}}, \hat{\boldsymbol{\Sigma}}). \tag{6.7}$$

There are then two simple alternatives for using the approximation.

1. Use the density directly to make independent proposals where the ith proposal is $\boldsymbol{\theta}'_i \sim N(\hat{\boldsymbol{\mu}}, \hat{\boldsymbol{\Sigma}})$. Because this is not a symmetric proposal the ratio $q(\boldsymbol{\theta}|\boldsymbol{\theta}')/q(\boldsymbol{\theta}'|\boldsymbol{\theta})$ is no longer identically 1, but the computation simply involves calculating the ratio of the multivariate normal density evaluated at the two parameter vector values.
2. Use a shrunken version of $\hat{\boldsymbol{\Sigma}}$ as the basis for proposing multivariate normal jumps in a random walk. So the ith proposal is $\boldsymbol{\theta}'_i \sim N(\boldsymbol{\theta}_{i-1}, \hat{\boldsymbol{\Sigma}}k^2)$. It turns out that for high dimensions $k = 2.4/\sqrt{d}$ is about optimal (see e.g. Gelman et al., 2013), although in any particular case some tuning is likely to be beneficial. In this case the probability densities of a move and the reverse move are equal, so the q ratio is 1.

Figure 6.9 illustrates the two approaches for a posterior shown as black contours. The left-hand panel shows option 1, in which proposals are generated directly from the normal approximation contoured in grey. The proposal is good in the centre of the distribution, but visits some of the tail regions very infrequently, relative to their posterior density. Metropolis Hastings compensates for this deficiency by leaving the chain stuck at such tail values for a long time, on the rare occasions that they are proposed. The black dot at the top right of the plot is such a tail value. What causes the stickiness of the chain is the ratio $q(\boldsymbol{\theta}|\boldsymbol{\theta}')/q(\boldsymbol{\theta}'|\boldsymbol{\theta})$. Reaching the point is highly improbable according to the proposal, so $q(\boldsymbol{\theta}|\boldsymbol{\theta}')$ (which actually does not depend on $\boldsymbol{\theta}'$ here) is tiny. In contrast $\boldsymbol{\theta}'$ is typically not in the far tails of the proposal, so that $q(\boldsymbol{\theta}'|\boldsymbol{\theta})$ is modest: hence the ratio is tiny, the MH acceptance probability is tiny, and it takes many iterations to leave the point. In consequence option 1 is only advisable when the normal approximation to the posterior is expected to be good.

The right panel of Figure 6.9 illustrates option 2: random walk updates based on a shrunken version of the covariance matrix estimate. Proposal densities are contoured in grey for two points: the open circle in the high posterior density region, and the black circle in the tail region. The proposal is reasonable in the high-density region. Similarly, the proposal has a reasonable chance of reaching the tail point in the first place and of leaving it again.

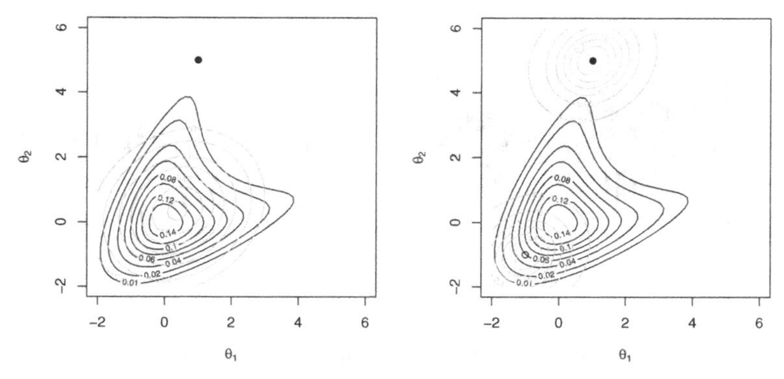

Figure 6.9 Simple proposals based on a multivariate normal (MVN) approximation of the posterior. The true posterior is contoured in black. Left: An MVN approximation to the posterior is contoured in grey, based on the empirical mean and covariance matrix of a sample from the posterior. Proposals could be made directly from this approximation, but then the probability of ever proposing the point shown as a black dot is very low, despite it having non-negligible posterior probability density. Hence when such a proposal *is* made, the chain tends to get stuck at this point for many iterations. Right: the density of a random walk proposal for the same point as grey contours, where the covariance of the proposal is a shrunken version of the covariance from the normal approximation. Clearly the random walk proposal has a better chance of reaching the point in the first place and of leaving it again. Also contoured in grey is the proposal density for the point marked by an open circle: the random walk proposal is also reasonable in the high posterior density region.

6.5.2 The curse of dimensionality

The attentive reader will have noticed that the random walk proposal discussed in the previous section requires that the proposal standard deviation is proportional to $1/\sqrt{d}$ where d is the dimension of $\boldsymbol{\theta}$. In other words, as dimension increases, the change proposed for each component of $\boldsymbol{\theta}$ has to be reduced in order to get optimal mixing. Figure 6.10 illustrates the inevitability of this effect by using Metropolis Hastings to sample from a $N(\mathbf{0}, \mathbf{I}_d)$ density for $d = 2$ and then for $d = 100$. A random walk proposal $\boldsymbol{\theta}'_{i+1} \sim N(\boldsymbol{\theta}_i, \mathbf{I}_d\sigma_p^2)$ is used, where σ_p is tuned to achieve the maximum effective sample size, separately for each d. Clearly there is no issue with correlation here, but mixing is still very slow for the relatively high dimensional problem.

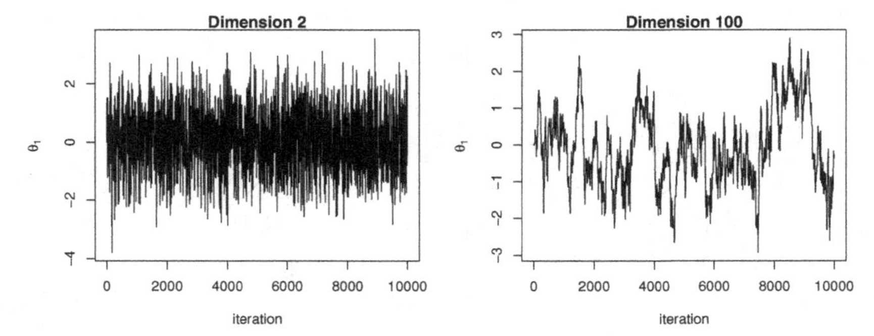

Figure 6.10 Metropolis Hastings sampling from $N(\mathbf{0}, \mathbf{I}_d)$ for $d = 2$ (left) and $d = 100$ (right). In both cases proposals were of the form $\boldsymbol{\theta}'_{i+1} \sim N(\boldsymbol{\theta}_i, \mathbf{I}_d\sigma_p)$. Different σ_p were used for $d = 2$ and $d = 100$, each tuned to give the largest possible effective sample size. Notice how mixing is much slower for the higher dimensional problem: a simple result of geometry.

This effect is geometrically inescapable when using symmetric random walk proposals. The fundamental problem is that, as the dimension increases, a symmetric random walk proposes ever fewer jumps that actually have increased posterior density, relative to the starting point. Figure 6.11 illustrates this dropoff when moving from $d = 1$ to $d = 2$ dimensions, considering the simple case in which the target posterior density is $N(\mathbf{0}, \mathbf{I}_d)$ and the proposal is based on independent $U(-\sqrt{d}, \sqrt{d})$ (approx.) increments for each element of θ. As d increases, this problem becomes ever more severe, especially near the centre of the distribution. Indeed considering such uniform proposals from points like the black blobs in Figure 6.11, it is easy to work out the probability of a proposal falling in the region of increased posterior probability. It is half the volume of an r-radius d-ball divided by the volume of a d-dimensional hyper cube of side length $2r$: $\pi^{d/2}/\{\Gamma(d/2+1)2^{d+1}\}$. This probability drops from 0.5 for $d = 1$ to less than 1% by $d = 8$.

The example used here to illustrate the issue is far from pathological. There is no correlation present, and the density is as well behaved as we could hope. In addition, we could transform any multivariate normal density to this case without loss of generality, and in the large sample limit many posteriors tend to multivariate normality.

6.5.3 Improved proposals based on approximate posterior normality

These basic issues have given rise to a great deal of work on adaptive Metropolis-Hastings schemes that use nonsymmetric proposals with an

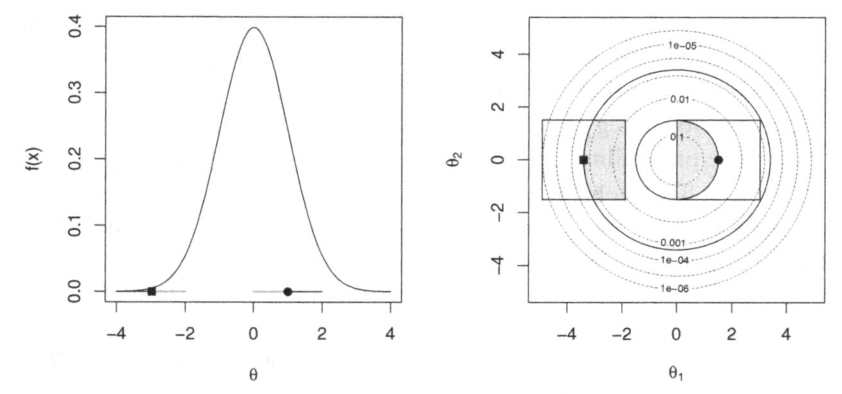

Figure 6.11 Symmetric random walk proposals become less probable with increasing dimension. Left: a one dimensional $N(0,1)$ p.d.f. About 68% of draws from $N(0,1)$ are more probable than the black blob, whereas 99.7% of draws are more probable than the black square. Consider proposing a symmetric uniformly distributed jump from the black blob that could just reach $\theta = 0$. The possible range of such a proposal is illustrated by the horizontal line through the blob: 50% of such proposals (shown in grey) have higher probability than the starting value, and 50% lower (shown in black). The same proportions apply if the proposal is used starting from the black square. Right: the situation in two dimensions when the density is $N(\mathbf{0}, \mathbf{I}_2)$. Again 68% of draws from $N(\mathbf{0}, \mathbf{I}_2)$ are more probable than the black blob, and 99.7% are more probable than the black circle. Again consider symmetric uniform proposals centred on the points and just capable of reaching 0, 0 for the black blob. The regions uniformly covered by the proposals are shown as squares centred on the two points. For a proposal to land in an area of increased density, it must be within the black contour passing through the starting point (i.e within the region shaded grey). Clearly for the black blob far fewer than 50% of proposals will end up in the grey region of increased probability. Further out in the tails, where the black square is, the chance is higher, but still less than 50%.

improved tendency to make moves that will increase the posterior, while taking relatively large steps. Most of this work is well beyond our scope here. Instead let us consider developing some simple improved schemes, based on the preceding insights.

The big advantage of the fixed multivariate normal proposal is that it tends to propose more points near the centre of the distribution than in the tails, but its disadvantage is that it may not visit poorly approximated tails often enough, so it gets stuck in them when it does visit. The advantage of the random walk is that it can get into the tails without getting stuck there, but does so by making many proposals into low-probability regions. An obvious hybrid strategy is to propose moves from a mixture distribution. With tunable probability γ propose from $N(\hat{\boldsymbol{\mu}}, \hat{\boldsymbol{\Sigma}})$ otherwise propose from $N(\boldsymbol{\theta}_i, \hat{\boldsymbol{\Sigma}}k^2)$. The probability density for the proposal and its reverse now have to be computed from the mixture distribution to compute the MH q ratio, but this is unproblematic.

A simpler alternative that directly addresses the tendency of symmetric random walks to propose improbable moves in high dimensions is to move the centre of such proposals from $\boldsymbol{\theta}_i$, in the direction of $\hat{\boldsymbol{\mu}}$. Defining $\|\boldsymbol{\theta} - \hat{\boldsymbol{\mu}}\|^2_{\hat{\Sigma}} = (\boldsymbol{\theta} - \hat{\boldsymbol{\mu}})^{\mathrm{T}}\hat{\boldsymbol{\Sigma}}^{-1}(\boldsymbol{\theta} - \hat{\boldsymbol{\mu}})$ and

$$\mathbf{m}(\boldsymbol{\theta}) = \begin{cases} \boldsymbol{\theta} - \gamma(\boldsymbol{\theta} - \hat{\boldsymbol{\mu}})/\|\boldsymbol{\theta} - \hat{\boldsymbol{\mu}}\|_{\hat{\Sigma}} & \|\boldsymbol{\theta} - \hat{\boldsymbol{\mu}}\|_{\hat{\Sigma}} > \gamma, \\ \boldsymbol{\theta} & \text{otherwise} \end{cases}$$

the proposal density becomes $N(\mathbf{m}(\boldsymbol{\theta}), \hat{\boldsymbol{\Sigma}}k^2)$. We must choose γ and can typically afford to increase k somewhat.

6.5.4 *Improved proposals for the algal population example*

The samplers constructed in Section 6.4 have rather disappointing performance, in terms of effective sample size for computational effort. 150,000 iterations still only gave an effective sample size of around 370 for n_0, and each of those updates required an accept/reject computation for each element of $\boldsymbol{\theta}$ and $\boldsymbol{n}$ separately. This section compares the simple improved updates discussed in the previous subsection based on the covariance matrix of the parameters according to the first run.

If the state vectors `n` for each iteration are stored as columns of a matrix `nn`, then the first step is to compute the mean and covariance matrix for the vector $\mathbf{b} = (\boldsymbol{\theta}^{\mathrm{T}}, \mathbf{n}^{\mathrm{T}})^{\mathrm{T}}$:

```
## tn is params and state, discarding burn-in...
tn <- rbind(th,nn)[,-(1:20000)]
mu <- rowMeans(tn) ## mu hat
V <- cov(t(tn)) ## Sigma hat
```

Before going further it is important to look at `pairs` plots of the rows of `tn` to see whether $N(\hat{\boldsymbol{\mu}}, \hat{\boldsymbol{\Sigma}})$ can be expected to capture anything useful about the posterior. In this case it can, so first consider using $N(\hat{\boldsymbol{\mu}}, \hat{\boldsymbol{\Sigma}})$ as a

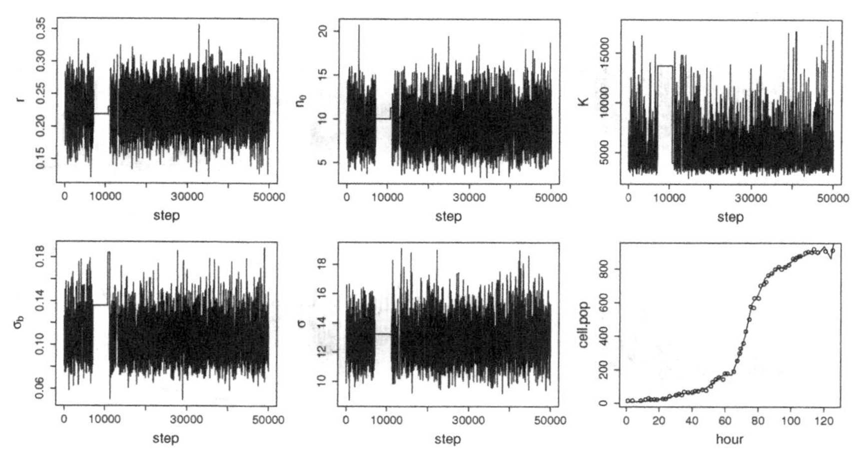

Figure 6.12 Algal model MCMC output equivalent to Figure 6.6, for 50000 iterations using a fixed multivariate normal proposal, based on the mean and covariance matrix from a pilot run. Notice how the chain becomes stuck where the posterior density is moderate but the proposal density is very low; it then takes a large number of iterations to become unstuck: see the discussion around Figure 6.9 for an explanation of this phenomenon.

fixed proposal. All proposals can be generated up front, using, for example `mvrnorm` from the `MASS` library in R:

```
library(MASS)
sp <- mvrnorm(n.mc,mu,V)
```

so `sp` contains one proposal per row. The q ratio in the MH acceptance probability requires the density of each proposal, and the following will evaluate the log of these for all rows of `sp`:

```
dmvn <- function(x,mu,V) {
## one vec in each col of x
  R <- chol(V)
  z <- forwardsolve(t(R),x-mu)
  -colSums(z^2)/2-sum(log(diag(R)))-log(2*pi)*length(mu)/2
}
lfsp <- dmvn(t(sp),mu,V)
```

Hence, if the chain is in state $\mathbf{b}$ = `sp[i,]` and the proposal is $\mathbf{b}'$ = `sp[j,]`, then `exp(lfsp[i]-lfsp[j])` gives $q(\mathbf{b}|\mathbf{b}')/q(\mathbf{b}'|\mathbf{b})$. Figure 6.12 shows output from the chain with this proposal. Notice the lengthy period in which the chain is stuck, as discussed in Section 6.5.1 and Figure 6.9.

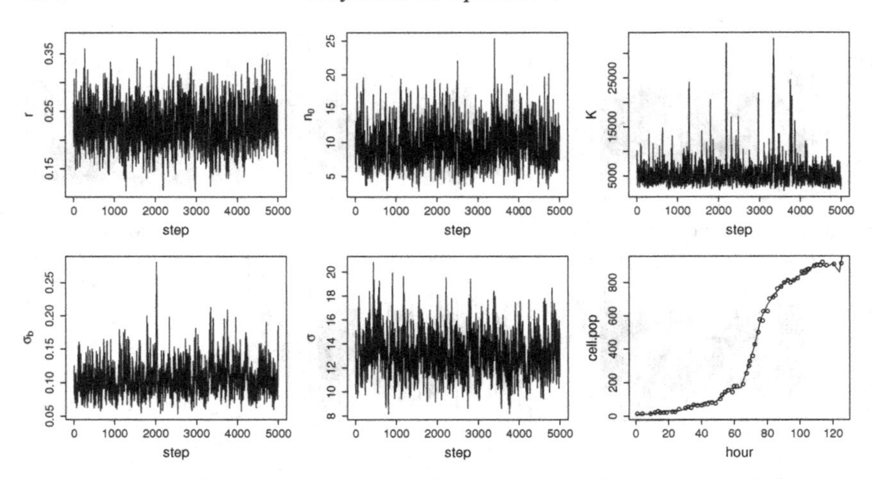

Figure 6.13 Algal model MCMC output equivalent to Figure 6.6, for 50000 iterations using a random walk proposal with correlated multivariate normal jumps, based on a shrunken version of the covariance matrix from a pilot run. Although much more efficient than the chain used for Figure 6.6, the curse of dimensionality discussed in Section 6.5.2 means that progress is still fairly slow.

Clearly, these results are not satisfactory. We definitely cannot discard the stuck section of chain, because the stuck section is the only thing ensuring that this region of the posterior is sampled in the correct proportion. But the facts that this section exists and that there is only one such section are clear indications that we have not run the chain for long enough to adequately sample this region of the posterior.

The random walk proposal in which $\mathbf{b}' \sim N(\mathbf{b}, \hat{\boldsymbol{\Sigma}}k^2)$ does not get stuck in this way. `sp <-mvrnorm(n.mc,rep(0,ncol(V)),V)` can be used to generate the jumps up front, so the proposals are `bp <-b + sp[i,]*k`. Now the q ratio is 1. Figure 6.13 shows results for such a chain. Minimum effective sample sizes of around 350 from 50000 iterations indicate that this approach is quite a bit more efficient than the original sampler of Section 6.4, which required 150,000 much more expensive steps to achieve the same. However the curse of dimensionality still leads to slow mixing here.

Of the alternatives in Section 6.5.3, the biased random walk gives better results than the mixture proposal for this problem and is also easier to implement, so let us consider it here. The densities required for the q ratio cannot be computed up front now, so we need to be able to compute them efficiently during the computation. This is easy if the Choleski decomposition of the covariance matrix is computed before the iteration, so the following implementation does this:

```
rwp <- mvrnorm(n.mc,rep(0,nrow(V)),V) ## used for jumps

dmvnr <- function(x,mu,R) {
## computes log density of x~N(mu,R'R)
  z <- forwardsolve(t(R),x-mu)
  -sum(z^2)/2 - sum(log(diag(R))) - log(2*pi)*length(mu)/2
}

R <- chol(V) ## R'R = V
th <- matrix(0,length(ll),n.mc)
theta <- ind <- 1:5
b0 <- b <- mu ## combined theta and n
th[,1] <-theta <- b[ind]; n <- b[-ind]

lf0 <- lfey(theta,n,y,li)
accept <- 0

gamma=.5; ## dist. to move prop. mu towards overall mean
k <- 2.4/sqrt(length(b))*1.2 ## jump scale

## compute first proposal mean vector, muw...
z <- forwardsolve(t(R),b-mu); dz <- sqrt(sum(z^2))
if (dz>gamma) muw <- b - (b-mu)*gamma/sqrt(sum(z^2)) else
              muw <- b

for (i in 2:n.mc) { ## mcmc loop
  muw.0 <- muw ## mean of current state
  b0 <- b
  b <- muw.0 + rwp[i,] * k ## proposal from N(muw.0,V*k^2)
  ## find mean of proposal starting from b...
  z <- forwardsolve(t(R),b-mu);dz <- sqrt(sum(z^2))
  if (dz>gamma) muw <- b-(b-mu)*gamma/sqrt(sum(z^2)) else
              muw <- b

  theta <- b[ind]; n <- b[-ind]
  lf1 <- lfey(theta,n,y,li)
  q.rat <- dmvnr(b0,muw,R*k)-dmvnr(b,muw.0,R*k)
  if (runif(1) < exp(lf1-lf0+q.rat)&&
                  sum(theta>ul|theta<ll) == 0) { ## accept
    accept <- accept + 1
    lf0 <- lf1
  } else { ## reject
    b <- b0
    muw <- muw.0
  }
  th[,i] <- theta
  if (i%%3000==0) cat(".")
} ## end of loop
```

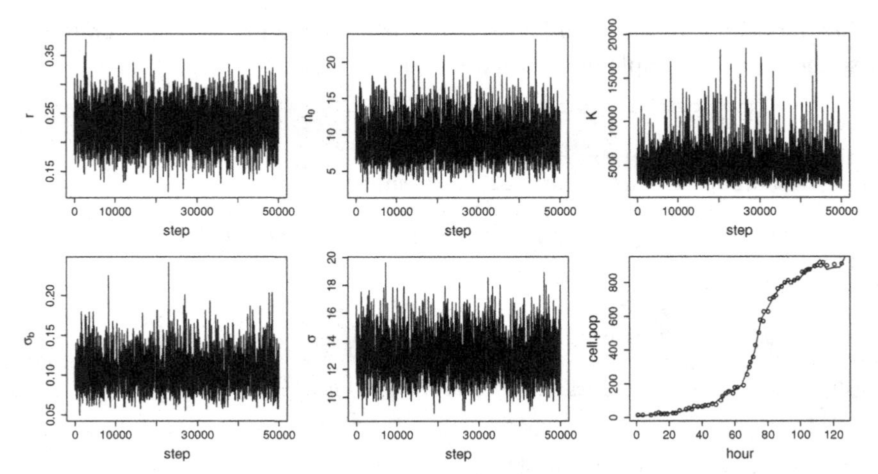

Figure 6.14 Algal model MCMC output equivalent to Figure 6.6, for 50000 iterations using a biased random walk proposal with correlated multivariate normal jumps, based on a shrunken version of the covariance matrix from a pilot run, and with a mean moved from the current state of the chain towards the overall mean from the pilot run. This is the most efficient of the samplers tried here.

This chain achieves a minimum effective sample size of 1200 in 50000 iterations: the better mixing can be seen in Figure 6.14. In terms of computer time per minimum effective sample size, this sampler is approximately 100 times more efficient than the original sampler in Section 6.4. Notice that it only worked because the posterior normality approximation was not too bad in this case: had the posterior been less amenable, then something more sophisticated would have been required.

6.6 Graphical models and automatic Gibbs sampling

The implementation of MCMC samplers is obviously rather time consuming, and the question of automating the construction of samplers arises. It turns out that automatic Gibbs sampling can be very successfully implemented for *Bayesian graphical models* in which the dependency structure between variables in the model can be represented by a *directed acyclic graph* (DAG). The basic trick is to break down simulation from a high-dimensional posterior into a sequence of Gibbs sampling steps of intrinsically low dimension.

The automation process is bound up with the model's DAG structure, so we need to explore the concepts here. Graphs already featured in Section 5.5.3, where computational graphs (examples of DAGs) were used in automatic differentiation. A directed graph consists of a set of *nodes* connected by directed edges: arrows. These arrows run from *parents* to *children*. Every variable in a graphical model is a node, and the key feature of such models is that the distribution of a variable/node is completely known if you know the values of all its parent nodes. The fact that the graphs are *acyclic* means that no node is its own ancestor: you cannot find a path through the graph that follows edges in the direction of the arrows and arrives back at the node you started from.

It is helpful to distinguish three types of node.

1. Stochastic nodes are variables with a distribution that depends stochastically on other nodes. They may be observed (i.e. correspond to data), or unobserved (parameters or random effects).
2. Deterministic nodes are nodes that are deterministic (logical) functions of other nodes. They cannot be observed.
3. Constants are fixed numbers and have no parents.

There are two types of arrows. Deterministic/logical relationships between nodes are usually shown as dashed arrows, whereas stochastic relationships are shown as solid arrows.

Figure 6.15 illustrates a portion of the DAG for the algal population model of Section 6.4, assuming that proper gamma(α_j, λ_j) priors have been specified for r, K, $1/\sigma_e^2$ and $1/\sigma^2$. The parentless nodes at the far left and right of the figure are the constants specifying the various gamma priors: actual numbers would have to be supplied here. The portion of the graph shown surrounds a data node y_τ whose time of observation lies between discrete update times t and $t{+}1$, so that its expected value is obtained by linear interpolation between nodes n_t and n_{t+1}. The fact that this linear interpolation is purely deterministic is the reason that the arrows from n_t and n_{t-1} to deterministic node μ_τ are dashed.

Now consider what makes graphical models convenient for automatic Gibbs sampling. Generically, let x_i denote the variable corresponding to the ith node of the graph. From the dependencies encoded in the graph it follows that the joint density of the (non constant) nodes is

$$f(\mathbf{x}) = \prod_i f(x_i|\text{parent}\{x_i\}), \tag{6.8}$$

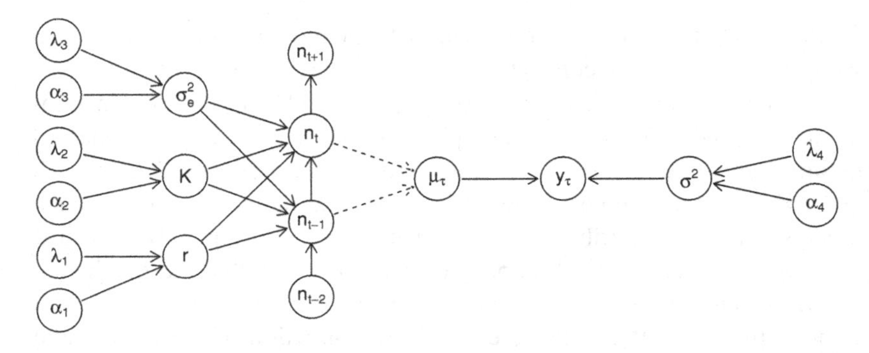

Figure 6.15 Part of the DAG for the algal model introduced in Section 6.4, assuming the priors introduced in Section 6.6. The figure focuses on a section of the graph relating to an observation at time τ lying between discrete simulation times t and $t+1$. Most of the edges connected to nodes n_{t-2} and n_{t+1} are not shown. Notice how the left and rightmost nodes, which are fixed constants defining priors, have no parents. Conversely, the observed data node y_t has no children, but this is a feature of this model, rather than being a requirement of the formalism.

where the product is over the non constant nodes. If this is not obvious, start with the childless (terminal) nodes and work back using the basic relations between conditional and joint densities covered in Section 1.4.2.

Gibbs sampling involves simulating from the full conditionals of all stochastic nodes other than those corresponding to data, which are fixed at their observed values. It turns out that these conditionals usually involve far fewer terms than the full joint density. From the definition of a conditional p.d.f.,

$$f(x_j|\mathbf{x}_{-j}) = \frac{f(\mathbf{x})}{\int f(\mathbf{x})dx_j} = \frac{\prod_i f(x_i|\text{parent}\{x_i\})}{\int \prod_i f(x_i|\text{parent}\{x_i\})dx_j},$$

but the only terms in the product that have to stay inside the integral are those that involve x_j: the conditional density of x_j given its parents, and the conditional densities of each child of x_j, given that child's parents. All other terms in (6.8) can be taken out of the integral and therefore cancel between the top and bottom of $f(x_j|\mathbf{x}_{-j})$. In short,

$$\begin{aligned} f(x_j|\mathbf{x}_{-j}) &= \frac{f(x_j|\text{parent}\{x_j\}) \prod_{i\in\text{child\{j\}}} f(x_i|\text{parent}\{x_i\})}{\int f(x_j|\text{parent}\{x_j\}) \prod_{i\in\text{child\{j\}}} f(x_i|\text{parent}\{x_i\})dx_j} \\ &\propto f(x_j|\text{parent}\{x_j\}) \prod_{i\in\text{child\{j\}}} f(x_i|\text{parent}\{x_i\}), \end{aligned}$$

so that, however complicated the model and corresponding DAG, $f(x_j|\mathbf{x}_{-j})$, required for the Gibbs update of x_j, depends only on the parent-conditional-densities of x_j and its children.

6.6.1 Building the samplers

The preceding discussion forms the basis for the automatic construction of Gibbs samplers. The model's DAG structure is used to identify the relatively small number of terms that play a part in each $f(x_j|\mathbf{x}_{-j})$, an attempt is made to identify the exact distribution for $f(x_j|\mathbf{x}_{-j})$, and when this is not possible a more costly general-purpose sampler is constructed. Identification of the exact distributions rests on known *conjugacy* relationships between distributions, while the ingenious method of *slice sampling* is often useful otherwise.

Conjugate distributions

Again consider,

$$f(x_j|\mathbf{x}_{-j}) \propto f(x_j|\text{parent}\{x_j\}) \prod_{i \in \text{child}\{j\}} f(x_i|\text{parent}\{x_i\}).$$

The right hand side has exactly the structure of a prior, $f(x_j|\text{parent}\{x_j\})$, for x_j, multiplied by a likelihood term for x_j, in which the children x_i play the role of data. This fact allows what is known about *conjugacy* of distributions to be exploited in automatically ascertaining the density that gives $f(x_j|\mathbf{x}_{-j})$.

If the prior and posterior distribution for some quantity are from the same family[7], for a given likelihood, then that distribution is said to be *conjugate* for that likelihood. We have already seen an example of this when constructing the simple Gibbs sampler in Section 6.2.8: there the normal distribution was shown to be conjugate for the mean of a normal likelihood term, while the gamma distribution was conjugate for the precision (reciprocal variance) of a normal. A whole library of such standard conjugacy results is known,[8] and can therefore be exploited in the automatic construction of Gibbs samplers.

[7] For example, a normal prior yields a normal posterior, or a gamma prior yields a gamma posterior. Of course, the parameters of the distributions change from prior to posterior.

[8] Bayesian statisticians had to have something to do between the reading of the Reverend Bayes' paper to the Royal Society of London in 1763, and the advent of computers cheap enough to make the Metropolis Hastings algorithm of 1953/1970 something usable for those with a smaller budget than the US atomic weapons program.

Slice sampling

When no convenient analytic form for a density $f(x_j|\mathbf{x}_{-j})$ can be obtained, then some other stochastic simulation method must be employed for that density. A beautifully simple approach is *slice sampling* (Neal, 2003). The basic observation is this: if we plot $kf(x)$ against x for any finite non-zero k, and then generate a coordinate x, y from a uniform density over the region bounded by $kf(x)$ and the x axis, then the resulting x value will be a draw from $f(x)$.

The problem, of course, is that generating directly from the required uniform density of x, y is no easier than generating from $f(x)$ itself. The simplicity arises when we consider a Gibbs update of x and y. Trivially

$$f(y|x) \sim U(0, kf(x))$$

while

$$f(x|y) \sim U(x : kf(x) \geq y),$$

so a Gibbs update would draw a y uniformly from the interval $(0, kf(x))$ and then draw x uniformly from the set of x values for which $kf(x) \geq y$ (the 'slice' of the technique's name). The only problem now is identifying the required set of x values. For a unimodal distribution, this set will constitute a single interval, which may be easy to locate, but for multimodal distributions several intervals may need to be identified. Of course, in practice it is only necessary to identify an interval or a set of intervals that bracket the required set: then we can generate uniformly on the bracketing interval(s) until we obtain an x such that $kf(x) \geq y$. If the bracketing interval(s) are too wide this will be inefficient, of course.

6.6.2 BUGS and JAGS

The most widely used software for statistics via automatically constructed Gibbs samplers is BUGS (Bayesian Updating via Gibbs Sampling), now followed by openBUGS, which has an R package interface `brugs`. BUGS established a simple language for the specification of graphical models, leading to other implementations including JAGS (Just Another Gibbs Sampler), with the R interface package `rjags`, which is covered here.

JAGS has to be installed as a standalone program and can be used as such, but it is very convenient to use it via `rjags`, which is the method covered here. `rjags` also provides easy integration with the `coda` package for convergence diagnostics. JAGS models are specified in a text file using a dialect of the BUGS language. The name of this file, together with a list providing the corresponding data, is supplied to the `jags.model` function, which calls JAGS itself to automatically generate a sampler, returned as an object of class `"jags"`. This object can then be used to generate samples using calls to `jags.samples` or the closely related `coda.samples` (depending on exactly what format you would like the data returned in).

Toy example

Recall the toy normal model example of Section 6.2.8 in which we have $n = 20$ observations $y_i \sim N(\mu, \phi)$, where $1/\phi \sim G(a, b)$ (a gamma random variable) and (independently) $\mu \sim N(c, d)$. a, b, c and d are constants. In graphical model terms each y_i, μ and $\tau = 1/\phi$ are stochastic nodes, whereas a, b, c and d are constant nodes: the graph has 26 nodes in total. The BUGS language is set up for convenient specification of each node, and their relationships (directed edges). Here is the contents of the file `norm.jags` coding up our toy model

```
model {
  for (i in 1:N) {
    y[i] ~ dnorm(mu,tau)
  }
  mu ~ dnorm(0.0, 0.01)
  tau ~ dgamma(0.05,0.005)
}
```

For anyone who has read this far, the language is very intuitive. The symbol `~` specifies a stochastic dependence (i.e. a $\rightarrow$ in the graph), with the BUGS/JAGS statement `y[i] ~ dnorm(mu,tau)` being exactly equivalent to the mathematical statement $y_i \sim N(\mu, 1/\tau)$. By default the normal distribution is parameterised in terms of its precision, rather than its variance. Notice the use of loops to deal with vectors. R programmers are used to avoiding loops for populating vectors, but here there is no problem: this code is going to be compiled by JAGS to produce a sampler, and the R efficiency concerns do not apply.

Here is the R code to get JAGS to build a sampler from R, given 20 observations in a vector, `y`:

```
library(rjags)
setwd("some/directory/somewhere")
jan <- jags.model("norm.jags",data=list(y=y,N=20))
```

Function `setwd` sets R's working directory to the location of `norm.jags`. `jags.model` then creates the sampler, setting the nodes identified in `data` to their observed values. JAGS counts `N` as a constant node in the model, so it reports that the model has 27 nodes, rather than the 26 counted before.

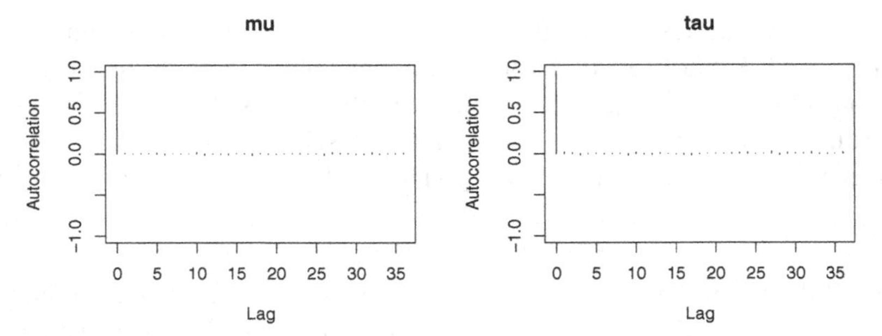

Figure 6.16 ACFs computed by `autocorr.plot` from the `coda` package for the toy model simulated using JAGS in Section 6.6.2. There is little autocorrelation here: to get more detail on the low correlations the `acfplot` function could be used.

The `jags.model` function also runs a number of adaptation iterations, tuning those component samplers that can be tuned to try to optimise their performance. The `n.adapt` argument controls the number of adaptation iterations and has a default value of 1000. At this stage `jan` contains a JAGS model object, which is ready to use to generate samples. Once built and initialized the sampler can be used:

```
> um <- jags.samples(jan,c("mu","tau"),n.iter=10000)
  |**************************************************| 100%
```

The second argument, `c("mu","tau")`, specifies the nodes that should be monitored at each step, and `n.iter` gives the number of steps (by setting argument `thin` to an integer greater than 1, we could monitor every `thin` steps). The results are stored in a two element list `um`, which could be used to produce a plot almost identical to Figure 6.2.

`rjags` can also create output in a manner convenient for use with the MCMC diagnostics package `coda`. Here is some code to do this, plot the chain ACFs shown in Figure 6.16 and compute effective sample sizes:

```
> er <- coda.samples(jan,c("mu","tau"),n.iter=10000)
  |**************************************************| 100%
> autocorr.plot(er)
> effectiveSize(er)
      mu       tau
 9616.071 10000.000
```

JAGS has a built-in function `dic.samples` for obtaining the DIC for a model. It uses a slightly different computation for p_D, which requires samples from two independent chains (see argument `n.chains` of `jags.model`). For example,

```
jan <- jags.model("norm.jags",data=list(y=y,N=20),
                                    n.chains=2)
dic.samples(jan,n.iter=10000)
```

The importance sampling method of Section 6.3.1 can also be used to estimate the marginal likelihood with JAGS. The sample for the posterior can be obtained as just shown, whereas the sample from the prior is obtainable by setting up and running the model with no data. The slightly inconvenient part is that $\log f(\mathbf{y}|\boldsymbol{\theta})$ has to be coded up again, externally to the JAGS model. But now let's move on to a less trivial example.

6.6.3 JAGS algal population example

The algal population growth model of Section 6.4 can easily be implemented in JAGS. The graph in this case has 874 nodes, and a portion of it is shown in Figure 6.15. The contents of the model specification file are as follows:

```
model {
  n[1] ~ dnorm(n0,tau)
  for (i in 2:M) {
    n[i] ~ dnorm(n[i-1] + r - exp(n[i-1]/K),tau)
  }
  for (i in 1:N) {
    mu[i] <- wm[i]*n[im[i]] + wp[i]*n[ip[i]]
    y[i] ~ dnorm(exp(mu[i]),tau0)
  }
  K ~ dgamma(1.0,.001)
  tau ~ dgamma(1.0,.1)
  r ~ dgamma(1.0,.1)
  n0 ~ dgamma(1.0,.1)
  tau0 ~ dgamma(1.0,.1)
}
```

Notice that there are two loops now. The first iterates the dynamic model for the log population for M steps. The second loop works through the N observed population nodes `y[i]`, relating them to the `n[i]`. The required linear interpolation is implemented via the deterministic nodes `mu[i]`, using the interpolation indices and weights as generated by the function `lint` defined in Section 6.4. The notation '<-' is equivalent to the dashed arrows in Figure 6.15.

The R code to use this model is a little more involved for this example, because we need to produce interpolation weights and an initial value for

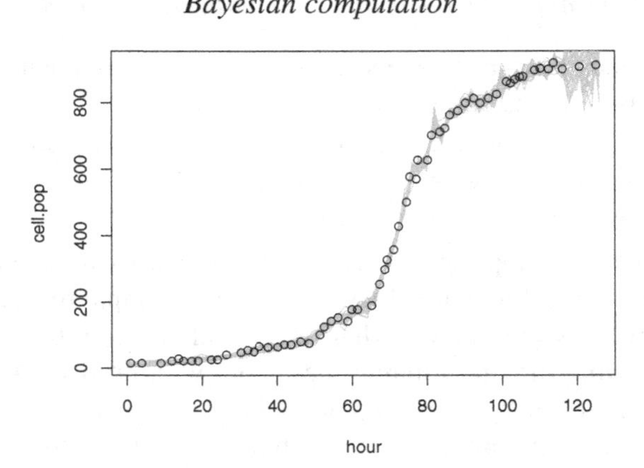

Figure 6.17 40 replicates of the underlying state $\exp(n)$ of the algal population model of section 6.6.3, as simulated using JAGS, shown as grey lines. The observed data are shown as open circles. Clearly the state is too variable towards the end of the data.

the state vector `n`. Without a reasonable initial value JAGS is unsuccessful at initialising this model.

```
library(rjags)
setwd("~location/of/model/file")
li <- lint(alg$hour,t0=0,dt=4)
dat <- list(y=alg$cell.pop,N=length(alg$cell.pop),M=li$m,
         ip=li$ip,im=li$im,wp=li$wp,wm=li$wm)
## initial state for n by linear interpolation of data...
ni <- log(c(alg$cell.pop[1],approx(alg$hour,alg$cell.pop,
         1:(li$m-2)*li$dt)$y,max(alg$cell.pop)))
jal <- jags.model("algae.jags",data=dat,inits=list(n=ni),
             n.adapt=10000)
ug <- coda.samples(jal,c("n0","r","K","tau0"),
              n.iter=40000,thin=10)
```

Here only every 10th sample has been stored in `ug`. According to `coda` the effective sample sizes are 1891, 559, 514 and 4000 for K, n_0, r and τ_0, respectively, and other diagnostic plots look reasonable. Let us look at the underlying state n, by having JAGS monitor it every 1000 iterations:

```
pop <- jags.samples(jal,c("n"),n.iter=40000,thin=1000)
plot(alg)
ts <- 0:(li$m-1)*li$dt
for (i in 1:40) lines(ts,exp(pop$n[,i,1]),col="grey")
with(alg,points(hour,cell.pop))
```

The results are shown in Figure 6.17: clearly the state is too variable at high population sizes, and the model would benefit from some modification.

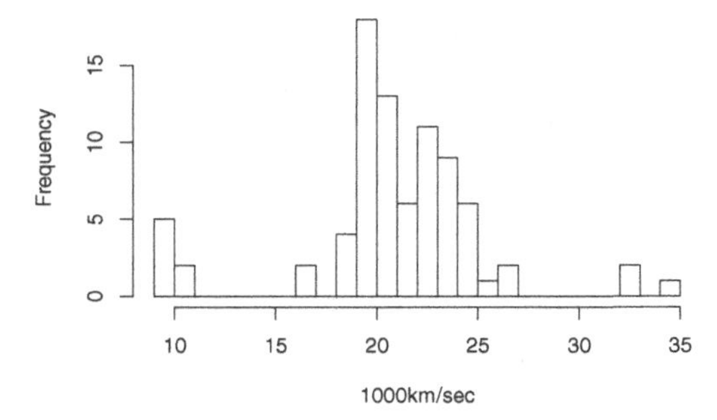

Figure 6.18 Histogram of the `galaxies` data from the `MASS` library (divided by 1000). The data show speeds of 82 galaxies, and the existence of several modes in the underlying distribution would be evidence for voids and large clusters in the far universe.

6.6.4 JAGS mixture model example

Figure 6.18 shows astronomical data on the speeds of 82 galaxies, where it is scientifically interesting to know whether the underlying distribution is multimodal (this is one of those classic datasets from the statistical literature). A popular approach to modelling such data is to use a mixture distribution; for example, to treat the data as coming from a mixture of normal densities. Letting y denote the speed of a randomly chosen galaxy in 1000kms^{-1} the p.d.f. might be

$$f(y) = \sum_{k=1}^{K} \alpha_k \phi(y; \mu_k, \sigma_k^2),$$

where $\phi(y; \mu_k, \sigma_k^2)$ denotes a normal p.d.f. with mean μ_k and variance σ_k^2, while positive mixture weights, α_k, sum to 1. A common approach to MCMC sampling from the posterior for the mixture parameters is to introduce auxiliary allocation variables, z_i say, which indicate from which component of the mixture each observation comes. Sampling is then from the posterior of the component mean and variances and the auxiliary variables. Notice the technical nuisance that we can permute the indices on μ_k, σ_k^2 without changing the model (the 'label switching problem'). In the one-dimensional case we could deal with this by re-parameterising, or just ignore it, but in the interests of keeping the pictures pretty I adopt the simple pragmatic devise of explicitly assigning one of the observations to each component (i.e. treat K of the z_i as known). Otherwise the z_i will be modelled as taking value k with probability α_k, where the α_k follow a

Dirichlet distribution (see Section A.2.4). Normal priors are used for the μ_k, and gamma priors are used for the $1/\sigma_k^2$.

The JAGS code is as follows, where z_i is `comp[i]`, `comp.tau[k]` is $1/\sigma_k^2$ and `comp.mu` is μ_k:

```
model {
  for (i in 1:N) {
    comp[i] ~ dcat(pc[1:K]) ## assign obs. to comp.s
    mu[i] <- comp.mu[comp[i]] ## pick out comp. mean
    tau[i] <- comp.tau[comp[i]] ## pick out comp. prec.
    y[i] ~ dnorm(mu[i],tau[i]) ## f(y|theta)
  }
  ## set up priors...
  pc[1:K] ~ ddirch(K.ones) ## Dirichlet prior
  for (i in 1:K) {
    comp.tau[i] ~ dgamma(1,.1)
    comp.mu[i] ~ dnorm(p.mean[i],1e-2)
  }
}
```

This can be utilised from R using something like the following, where a three component mixture is considered:

```
library(MASS);library(rjags)
y <- galaxies/1000 ## note y in ascending order
K <- 3;p.mean <- c(10,21,34) ## setting prior means & K
## fix obs closest to prior means to components...
comp <- rep(NA,N); comp[1] <- 1;comp[N] <- K
if (K>2) for (j in 2:(K-1)) {
  abs(y-p.mean[j])->zz; comp[which(zz==min(zz))] <- j
}
n.sim <- 20000
jam <- jags.model("mixture.jags",data=list(y=y,N=N,K=K,
       K.ones=rep(1,K),p.mean=p.mean,comp=comp),n.chains=1)
um <- jags.samples(jam,c("comp.mu","comp.tau"),
      n.iter=n.sim,thin=10)
```

Output from the chains is shown in Figure 6.19.

How many components should the mixture contain? The priors used in this case are vague and fairly arbitrary, so basing inference on posterior model probabilities or Bayes factors does not seem justified. The presence of the z_i allocation variables also makes it improbable that we are in the regime where the DIC is well justified. If we were to reformulate, and use maximum likelihood estimation directly with the likelihood based on the mixture distribution, then the BIC might be a possibility (provided we carefully check that the label switching problem can indeed be eliminated by a re-parameterisation compatible with the MLEs). However, a simpler approach based on predictive posterior simulation makes more scientific sense for this simple univariate situation.

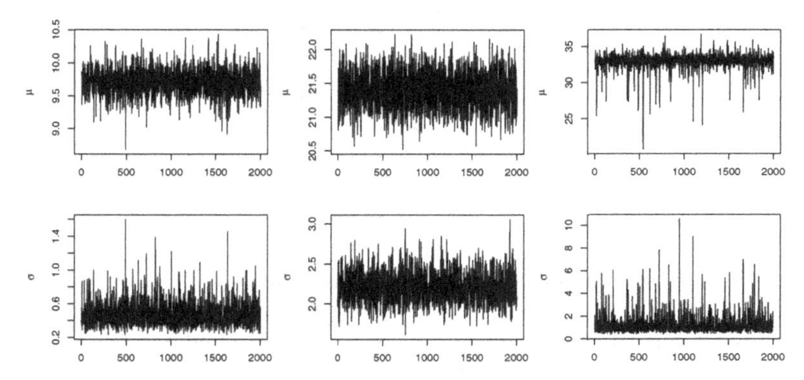

Figure 6.19 Chains for means and standard deviations of a three component mixture model for the galaxy data, simulated using JAGS, as discussed in Section 6.6.4.

Essentially we would like to choose the model most likely to produce data like the data we actually observe, and the obvious way to check this is to simulate new data given the posterior distributions of the model parameters. This is easy within JAGS. We simply add further nodes to the model to allow replicate data to be simulated. Here is the snippet of code to add to our original JAGS code to do this:

```
for (i in 1:N) {
  compp[i] ~ dcat(pc[1:K]) ## assign obs to comps
  mup[i] <- comp.mu[compp[i]] ## pick out comp mean
  taup[i] <- comp.tau[compp[i]] ## pick out comp prec
  yp[i] ~ dnorm(mup[i],taup[i]) ## pred. of y~f(y|theta)
}
```

So `yp` now contains new data simulated according to the posterior distribution of the model parameters (note that the auxilliary variables are drawn anew and not from their posterior). Having set this model up, exactly as before, we would sample from it, monitoring `yp`

```
um <- jags.samples(jam,c("yp"),n.iter=n.sim,thin=10)
```

To compare the posterior predictive distribution to that of the actual data we could look at QQ-plots, `plot(q,sort(y))`, based on quantile estimates `q <-quantile(um$yp[,,1],((0:81)+.5)/82)`. Figure 6.20 shows such plots for $K = 2$ to 4. More formally we could use the posterior predictive sample in `yp` to form the empirical c.d.f. of y, and then compute $u_i = \hat{F}^{-1}(y_i)$, which should be indistinguishable from $U(0,1)$ if the model is correct. Here is some code to compute such u_i and to test them for unifomity using a standard Kolmogorov-Smirnov test:

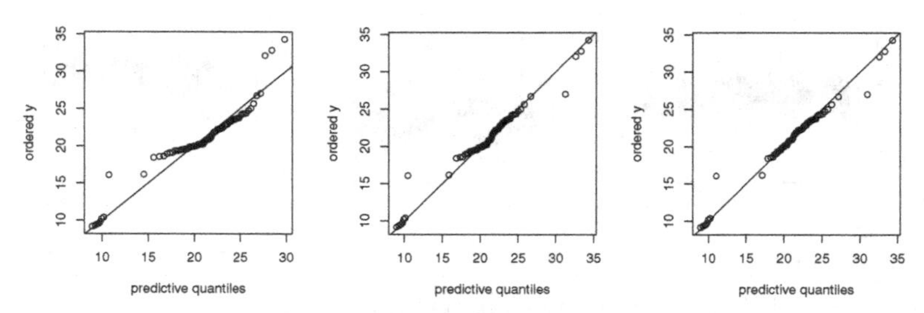

Figure 6.20 QQ-plots for galaxy data, when theoretical quantiles are generated by posterior simulation for K equal to 2 (left), 3 (middle) and 4 (right). Plots for $K > 5$ look very similar to the right panel.

```
n <- length(as.numeric(um$yp[,,1]))
uq <- (1:n-0.5)/n
u <- approx(sort(um$yp[,,1]),uq,y)$y
ks.test(u,"punif")
```

The resulting p-value of 0.32 provides no evidence that the galaxy data are not from a three component mixture, although the corresponding QQ-plot shows some deviation from a straight line. A two component mixture gives a quite ugly QQ-plot and a lower p-value, but still not into the firmly rejectable range. By four components the p-value is over 0.98 and the QQ-plot pretty much a straight line. Five components yields a marginally higher p-value, after which it starts to decline again. So with even the slightest preference for simplicity we would select the four component mixture, although it is hard to rule out a lower K given these data. See Gelman et al. (2013) for much more on checking via posterior model simulation.

6.6.5 JAGS urchin growth example

Now consider the urchin growth model of Section 5.4, but with vague gamma priors on the variance parameters and normal priors on mean parameters. Here is the JAGS model specification file (`urchin.jags`), where because the distributional assumption is on the square root of the observed volume, we need a `data` section in the model specification in order to implement the square root transformation:[9]

```
data {
for (i in 1:N) { rv[i] <- sqrt(v[i])}
}
```

[9] This is slightly different from the way BUGS deals with data transformation.

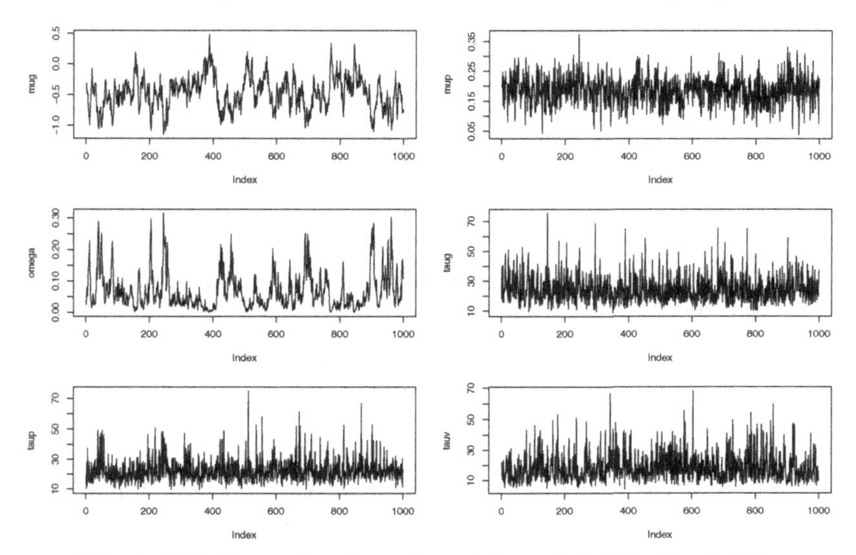

Figure 6.21 Trace plots from the urchin model of Section 6.6.5, simulated using JAGS. There is clear negative correlation between `omega` and `mug`, with consequent slow mixing of these chains.

```
model {
  for (i in 1:N) {
    p[i] ~ dlnorm(mup,taup)
    g[i] ~ dlnorm(mug,taug)
    am[i] <- log(p[i]/(g[i]*omega))/g[i]
    murv[i] <- (a[i] < am[i])*sqrt(omega*exp(g[i]*a[i])) +
        (a[i] >= am[i])*sqrt(p[i]/g[i] + p[i]*(a[i]-am[i]))
    rv[i] ~ dnorm(murv[i],tauv)
  }
  tauv ~ dgamma(1.0,.1)
  taup ~ dgamma(1.0,.1)
  taug ~ dgamma(1.0,.1)
  mup ~ dnorm(0,0.0001)
  mug ~ dnorm(0,0.0001)
  omega ~ dgamma(1.0,.1)
}
```

Note that `log(p[i]) ~ dnorm(mup,taup)` cannot be used in place of `p[i] ~ dlnorm(mup,taup)`, and `TRUE/FALSE` are interpreted as `1/0` in arithmetic expressions. Here is some code to set up and simulate from this model, assuming data in dataframe `uv`:

```
N <- nrow(uv)
jan <- jags.model("urchin.jags",
      data=list(v=uv$vol,a=uv$age,N=N))
```

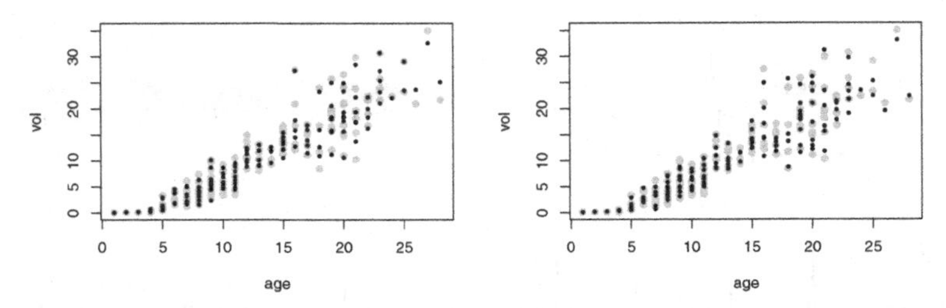

Figure 6.22 Two draws from the posterior distribution of the urchin volumes, shown as black dots overlaid on the data shown as grey circles.

```
um <- jags.samples(jan,c("mup","mug","taup","taug",
        "omega","tauv"),n.iter=100000,thin=100)
```

Mixing is slow for `omega` and `mug`, which appear to have high posterior correlation (correlation coefficient -0.9): see the trace plots in Figure 6.21. Now let us look at the predicted urchin volumes for two draws from the model, overlaid on the observed volumes:

```
rn <- jags.samples(jan,c("murv"),n.iter=1001,thin=1000)
par(mfrow=c(1,2))
plot(uv,col="grey",pch=19);
points(uv$age,rn$murv[,1,]^2,pch=19,cex=.5)
plot(uv,col="grey",pch=19);
points(uv$age,rn$murv[,2,]^2,pch=19,cex=.5)
```

The results are shown in Figure 6.22 and look reasonably plausible. It would also be worth simulating volumes from the posterior distribution of the model parameters by adding a replicate loop into the JAGS code, which is like the existing loop but with `p`, `g`, `am`, `murv` and `rv` replaced by predictive versions `pp`, `gp`, `amp`, `murvp` and `rvp`, respectively. `rvp` could then be monitored and compared to the original root volume data.

Exercises

6.1 The `nhtemp` data introduced in Section 2.1 can be modelled using the t_α based model given as the second example in that section.

a. Write a Metropolis Hastings sampler to simulate from the posterior of μ, σ and α, assuming improper uniform priors on μ and σ, but a proper geometric prior with $p = 0.05$ for $\alpha - 1$.
b. Check that the chains have converged.
c. Check the feasibility of the model with the aid of simulated new data from the posterior.

d. Produce an alternative model in which μ_i increases linearly with year, sample from the posterior distribution of the parameters in this case, and, by producing an appropriate credible interval, test whether there is evidence that average temperature is changing over the years.
e. Compare DIC values for the two versions of the model, and check whether they imply the same conclusion as the credible interval in this case.

6.2 Repeat the analysis from question 6.1 using JAGS.

6.3 Produce code to reproduce Figure 6.10 from Section 6.5.2, and investigate the use of the final improved proposal scheme of Section 6.5.3 using this example (of course in this case a static multivariate normal proposal would be optimal!)

6.4 Use JAGS to simulate from the posterior of the parameters of the bone marrow survival model given in Example 4 of Section 2.1. Use vague priors.

6.5 Model (6.6) in Section 6.4 can produce highly nonlinear 'chaotic' dynamics for sufficiently large values of r.

a. Simulate a time series of 50 data points from (6.6), using $r = 3.8$, $K = 20$, $N_0 = 10$, and assuming that what is actually observed is not N_t, but rather $Y_t \sim \text{Poi}(N_t)$.
b. Write code to simulate from the posterior of the model parameters, given the simulated y_t, bearing in mind that the interpolation required in Section 6.4 is not needed here, although it is still sensible to work in terms of $n_t = \log N_t$ rather than directly in terms of e_t.
c. Once the sampler is working, try writing a sampler that works in terms of e_t rather than n_t, and try to work out why it mixes so badly.

6.6 Produce a better model of the `geyser` data from the `MASS` library, introduced in Exercise 5.6, using both `waiting` times and `duration` of the eruptions. Sample from the posterior density of its parameters.

7

Linear models

This book has focused on general statistical methods that can be used with non standard models for statistical inference. This generality comes at the price of approximation, either through resort to large sample theory, in the case of most applications of maximum likelihood estimation, or use of stochastic simulation (or Laplace approximation) in the case of most Bayesian analyses. There is, however, one class of widely used statistical models for which inference, given the model, does not rely on approximation. These are *linear models*, and this chapter briefly covers their general theory and use.

A linear model is one in which a *response* vector $\mathbf{y}$ is linear in some parameters $\boldsymbol{\beta}$ and some zero mean random errors $\boldsymbol{\epsilon}$, so that

$$\mathbf{y} = \mathbf{X}\boldsymbol{\beta} + \boldsymbol{\epsilon}.$$

The *model matrix* $\mathbf{X}$ is determined by some known *predictor* variables (also known as *covariates*[1]), observed along with each response observation y_i. Usually the elements of $\boldsymbol{\epsilon}$ are assumed to be mutually independent with constant variance σ^2. For the purposes of finding confidence intervals and testing hypotheses for $\boldsymbol{\beta}$, the ϵ_i are also assumed to have a normal distribution.

Two types of predictor variable form the basic ingredients of $\mathbf{X}$.

1. *Metric* predictor variables are measurements of some quantity that may help to predict the value of the response. For example, if the response is the blood pressure of patients in a clinical trial, then age, fat mass and height are potential metric predictor variables.
2. *Factor* variables are labels that serve to categorize the response measurements into groups, which may have different expected values. Continuing the blood pressure example, factor variables might be sex and

[1] Response and predictor variables are sometimes known as 'dependent' and 'independent' variables, a particularly confusing terminology, not used here.

drug treatment received (drug A, drug B or placebo, for example). Somewhat confusingly, the groups of a factor variable are referred to as *levels* although the groups generally have no natural ordering, and even if they do, the model structure ignores it.

To understand the construction of $\mathbf{X}$ it helps to consider an example. Suppose that along with y_i we have metric predictor variables x_i and z_i and factor variable g_i, which contains labels dividing y_i into three groups. Suppose further that we believe the following model to be appropriate:

$$y_i = \gamma_{g_i} + \alpha_1 x_i + \alpha_2 z_i + \alpha_3 z_i^2 + \alpha_4 z_i x_i + \epsilon_i, \quad i = 1, \ldots, n,$$

where there is a different γ parameter for each of the three levels of g_i. Collecting the γ and α parameters into one vector, $\boldsymbol{\beta}$, we can rewrite the model in matrix-vector form as

$$\begin{bmatrix} y_1 \\ y_2 \\ y_3 \\ y_4 \\ \cdot \\ \cdot \\ \cdot \\ \cdot \\ y_n \end{bmatrix} = \begin{bmatrix} 1 & 0 & 0 & x_1 & z_1 & z_1^2 & z_1 x_1 \\ 1 & 0 & 0 & x_2 & z_2 & z_2^2 & z_2 x_2 \\ 1 & 0 & 0 & x_3 & z_3 & z_3^2 & z_3 x_3 \\ 0 & 1 & 0 & x_4 & z_4 & z_4^2 & z_4 x_4 \\ \cdot & \cdot & \cdot & \cdot & \cdot & \cdot & \cdot \\ \cdot & \cdot & \cdot & \cdot & \cdot & \cdot & \cdot \\ \cdot & \cdot & \cdot & \cdot & \cdot & \cdot & \cdot \\ \cdot & \cdot & \cdot & \cdot & \cdot & \cdot & \cdot \\ 0 & 0 & 1 & x_n & z_n & z_n^2 & z_n x_n \end{bmatrix} \begin{bmatrix} \gamma_1 \\ \gamma_2 \\ \gamma_3 \\ \alpha_1 \\ \alpha_2 \\ \alpha_3 \\ \alpha_4 \end{bmatrix} + \begin{bmatrix} \epsilon_1 \\ \epsilon_2 \\ \epsilon_3 \\ \epsilon_4 \\ \cdot \\ \cdot \\ \cdot \\ \cdot \\ \epsilon_n \end{bmatrix}$$

where y_1 - y_3 are in group 1, y_4 is in group 2, and y_n is in group 3 of the factor g. Notice how the factor levels/groups each get a dummy indicator column in the model matrix, with elements showing whether the corresponding y_i belongs to the group or not. Notice also how the metric variables can enter the model nonlinearly: the model is linear in the parameters and error term, but not necessarily in the predictors.

7.1 The theory of linear models

This section shows how the parameters, $\boldsymbol{\beta}$, of the linear model

$$\boldsymbol{\mu} = \mathbf{X}\boldsymbol{\beta}, \quad \mathbf{y} \sim N(\boldsymbol{\mu}, \mathbf{I}_n \sigma^2) \tag{7.1}$$

can be estimated by least squares. It is assumed that $\mathbf{X}$ is a matrix, with n rows, p columns and rank p ($n > p$). It is also shown that the resulting estimator, $\hat{\boldsymbol{\beta}}$, is unbiased and that, given the normality of the data, $\hat{\boldsymbol{\beta}} \sim N(\boldsymbol{\beta}, (\mathbf{X}^{\mathrm{T}}\mathbf{X})^{-1}\sigma^2)$. Results are also derived for setting confidence limits on parameters and for testing hypotheses about parameters: in particular the hypothesis that several elements of $\boldsymbol{\beta}$ are simultaneously zero.

In this section it is important not to confuse the *length* of a vector with its *dimension*. For example $(1, 1, 1)^{\mathrm{T}}$ has dimension 3 and length $\sqrt{3}$. Also note that no distinction has been made notationally between random variables and particular observations of those random variables: it is usually clear from the context which is meant.

7.1.1 Least squares estimation of $\boldsymbol{\beta}$

Point estimates of the linear model parameters, $\boldsymbol{\beta}$, can be obtained by the method of least squares; that is, by minimising the residual sum of squares

$$\mathcal{S} = \sum_{i=1}^{n} (y_i - \mu_i)^2,$$

with respect to $\boldsymbol{\beta}$, where $\boldsymbol{\mu} = \mathbf{X}\boldsymbol{\beta}$. This fitting objective follows directly from the log likelihood for the model, but even without the assumption of normality, the *Gauss-Markov theorem* says that minimising $\mathcal{S}$ w.r.t. $\boldsymbol{\beta}$ will produce the minimum variance linear unbiased estimator of $\boldsymbol{\beta}$.

To use least squares with a linear model, written in general matrix-vector form, first recall the link between the Euclidean length of a vector and the sum of squares of its elements. If $\mathbf{v}$ is any vector of dimension, n, then $\|\mathbf{v}\|^2 \equiv \mathbf{v}^{\mathrm{T}}\mathbf{v} \equiv \sum_{i=1}^{n} v_i^2$. Hence

$$\mathcal{S} = \|\mathbf{y} - \boldsymbol{\mu}\|^2 = \|\mathbf{y} - \mathbf{X}\boldsymbol{\beta}\|^2.$$

Since $\mathcal{S}$ is simply the squared (Euclidian) length of the vector $\mathbf{y} - \mathbf{X}\boldsymbol{\beta}$, its value will be unchanged if $\mathbf{y} - \mathbf{X}\boldsymbol{\beta}$ is rotated or reflected. This observation is the basis for a practical method for finding $\hat{\boldsymbol{\beta}}$ and for developing the distributional results required to use linear models.

Specifically, as with any real matrix, $\mathbf{X}$ can always be decomposed

$$\mathbf{X} = \mathbf{Q}\begin{bmatrix} \mathbf{R} \\ \mathbf{0} \end{bmatrix} = \mathbf{Q}_f\mathbf{R}, \tag{7.2}$$

where $\mathbf{R}$ is a $p \times p$ upper triangular matrix,[2] and $\mathbf{Q}$ is an $n \times n$ orthogonal matrix, the first p columns of which form $\mathbf{Q}_f$. Recall that orthogonal matrices rotate/reflect vectors, but do not change their length. Orthogonality

[2] That is, $R_{i,j} = 0$ if $i > j$. See also Section B.5.

also means that $\mathbf{Q}\mathbf{Q}^{\mathrm{T}} = \mathbf{Q}^{\mathrm{T}}\mathbf{Q} = \mathbf{I}_n$. Multiplying $\mathbf{y} - \mathbf{X}\boldsymbol{\beta}$ by $\mathbf{Q}^{\mathrm{T}}$ implies that

$$\|\mathbf{y} - \mathbf{X}\boldsymbol{\beta}\|^2 = \|\mathbf{Q}^{\mathrm{T}}\mathbf{y} - \mathbf{Q}^{\mathrm{T}}\mathbf{X}\boldsymbol{\beta}\|^2 = \left\| \mathbf{Q}^{\mathrm{T}}\mathbf{y} - \left[\begin{array}{c} \mathbf{R} \\ \mathbf{0} \end{array} \right] \boldsymbol{\beta} \right\|^2 .$$

Defining p vector $\mathbf{f}$ and $n - p$ vector $\mathbf{r}$ so that $\left[\begin{array}{c} \mathbf{f} \\ \mathbf{r} \end{array} \right] \equiv \mathbf{Q}^{\mathrm{T}}\mathbf{y}$, yields[3]

$$\|\mathbf{y} - \mathbf{X}\boldsymbol{\beta}\|^2 = \left\| \left[\begin{array}{c} \mathbf{f} \\ \mathbf{r} \end{array} \right] - \left[\begin{array}{c} \mathbf{R} \\ \mathbf{0} \end{array} \right] \boldsymbol{\beta} \right\|^2 = \|\mathbf{f} - \mathbf{R}\boldsymbol{\beta}\|^2 + \|\mathbf{r}\|^2 .$$

The length of $\mathbf{r}$ does not depend on $\boldsymbol{\beta}$ and $\|\mathbf{f} - \mathbf{R}\boldsymbol{\beta}\|^2$ can be reduced to zero by choosing $\boldsymbol{\beta}$ so that $\mathbf{R}\boldsymbol{\beta}$ equals $\mathbf{f}$. Hence,

$$\hat{\boldsymbol{\beta}} = \mathbf{R}^{-1}\mathbf{f} \tag{7.3}$$

is the least squares estimator of $\boldsymbol{\beta}$. Notice that $\|\mathbf{r}\|^2 = \|\mathbf{y} - \mathbf{X}\hat{\boldsymbol{\beta}}\|^2$, the residual sum of squares for the model fit.

7.1.2 The distribution of $\hat{\beta}$

The distribution of the estimator, $\hat{\boldsymbol{\beta}}$, follows from that of $\mathbf{Q}^{\mathrm{T}}\mathbf{y}$. Multivariate normality of $\mathbf{Q}^{\mathrm{T}}\mathbf{y}$ follows from that of $\mathbf{y}$, and since the covariance matrix of $\mathbf{y}$ is $\mathbf{I}_n\sigma^2$, the covariance matrix of $\mathbf{Q}^{\mathrm{T}}\mathbf{y}$ is

$$\mathbf{V}_{\mathbf{Q}^{\mathrm{T}}\mathbf{y}} = \mathbf{Q}^{\mathrm{T}}\mathbf{I}_n\mathbf{Q}\sigma^2 = \mathbf{I}_n\sigma^2 .$$

Furthermore,

$$E\left[\begin{array}{c} \mathbf{f} \\ \mathbf{r} \end{array} \right] = E(\mathbf{Q}^{\mathrm{T}}\mathbf{y}) = \mathbf{Q}^{\mathrm{T}}\mathbf{X}\boldsymbol{\beta} = \left[\begin{array}{c} \mathbf{R} \\ \mathbf{0} \end{array} \right] \boldsymbol{\beta}$$
$$\Rightarrow E(\mathbf{f}) = \mathbf{R}\boldsymbol{\beta} \text{ and } E(\mathbf{r}) = \mathbf{0}.$$

So we have that

$$\mathbf{f} \sim N(\mathbf{R}\boldsymbol{\beta}, \mathbf{I}_p\sigma^2) \text{ and } \mathbf{r} \sim N(\mathbf{0}, \mathbf{I}_{n-p}\sigma^2)$$

with both vectors independent of each other.

Turning to the properties of $\hat{\boldsymbol{\beta}}$ itself, unbiasedness follows immediately:

$$E(\hat{\boldsymbol{\beta}}) = \mathbf{R}^{-1}E(\mathbf{f}) = \mathbf{R}^{-1}\mathbf{R}\boldsymbol{\beta} = \boldsymbol{\beta}.$$

[3] If the final equality is not obvious recall that $\|\mathbf{x}\|^2 = \sum_i x_i^2$, so if $\mathbf{x} = \left[\begin{array}{c} \mathbf{v} \\ \mathbf{w} \end{array} \right]$, $\|\mathbf{x}\|^2 = \sum_i v_i^2 + \sum_i w_i^2 = \|\mathbf{v}\|^2 + \|\mathbf{w}\|^2$.

Since the covariance matrix of $\mathbf{f}$ is $\mathbf{I}_p\sigma^2$, it also follows from (1.5) in Section 1.5.1 that the covariance matrix of $\hat{\boldsymbol{\beta}}$ is

$$\mathbf{V}_{\hat{\beta}} = \mathbf{R}^{-1}\mathbf{I}_p\mathbf{R}^{-\mathrm{T}}\sigma^2 = \mathbf{R}^{-1}\mathbf{R}^{-\mathrm{T}}\sigma^2. \tag{7.4}$$

Furthermore, since $\hat{\boldsymbol{\beta}}$ is just a linear transformation of the normal random vector $\mathbf{f}$, it must have a multivariate normal distribution:

$$\hat{\boldsymbol{\beta}} \sim N(\boldsymbol{\beta}, \mathbf{V}_{\hat{\beta}}).$$

This result is not usually directly useful for making inferences about $\boldsymbol{\beta}$, because σ^2 is generally unknown and must be estimated, thereby introducing an extra component of variability that should be accounted for.

7.1.3 $(\hat{\beta}_i - \beta_i)/\hat{\sigma}_{\hat{\beta}_i} \sim t_{n-p}$

This section derives a result that *is* generally useful for testing hypotheses about individual β_i, as well as for finding confidence intervals for β_i. Since the $n-p$ elements of $\mathbf{r}$ are i.i.d. $N(0, \sigma^2)$ random variables,

$$\frac{1}{\sigma^2}\|\mathbf{r}\|^2 = \frac{1}{\sigma^2}\sum_{i=1}^{n-p} r_i^2 \sim \chi^2_{n-p}$$

(see Section A.1.2). The mean of a χ^2_{n-p} r.v. is $n-p$, so this result is sufficient (but not necessary) to imply that

$$\hat{\sigma}^2 = \|\mathbf{r}\|^2/(n-p) \tag{7.5}$$

is an unbiased estimator of σ^2. The independence of the elements of $\mathbf{r}$ and $\mathbf{f}$ also implies that $\hat{\boldsymbol{\beta}}$ and $\hat{\sigma}^2$ are independent.[4]

Now consider a single-parameter estimator, $\hat{\beta}_i$, with standard deviation, $\sigma_{\hat{\beta}_i}$, given by the square root of element i, i of $\mathbf{V}_{\hat{\beta}}$. An unbiased estimator of $\mathbf{V}_{\hat{\beta}}$ is $\hat{\mathbf{V}}_{\hat{\beta}} = \mathbf{V}_{\hat{\beta}}\hat{\sigma}^2/\sigma^2 = \mathbf{R}^{-1}\mathbf{R}^{-\mathrm{T}}\hat{\sigma}^2$, so an estimator, $\hat{\sigma}_{\hat{\beta}_i}$, is given by the square root of element i, i of $\hat{\mathbf{V}}_{\hat{\beta}}$, and it is clear that $\hat{\sigma}_{\hat{\beta}_i} = \sigma_{\hat{\beta}_i}\hat{\sigma}/\sigma$. Hence, using Section A.1.3,

$$\frac{\hat{\beta}_i - \beta_i}{\hat{\sigma}_{\hat{\beta}_i}} = \frac{\hat{\beta}_i - \beta_i}{\sigma_{\hat{\beta}_i}\hat{\sigma}/\sigma} = \frac{(\hat{\beta}_i - \beta_i)/\sigma_{\hat{\beta}_i}}{\sqrt{\frac{1}{\sigma^2}\|\mathbf{r}\|^2/(n-p)}} \sim \frac{N(0,1)}{\sqrt{\chi^2_{n-p}/(n-p)}} \sim t_{n-p} \tag{7.6}$$

[4] Recall that $\|\mathbf{r}\|^2 = \|\mathbf{y} - \mathbf{X}\hat{\boldsymbol{\beta}}\|^2$.

(where the independence of $\hat{\beta}_i$ and $\hat{\sigma}^2$ has been used). This result enables confidence intervals for β_i to be found and is the basis for hypothesis tests about individual β_is (for example, $\text{H}_0 : \beta_i = 0$).

7.1.4 F-ratio results

It is also of interest to obtain distributional results for testing, for example, the simultaneous equality to zero of several model parameters. Such tests are particularly useful for making inferences about factor variables and their interactions, because each factor (or interaction) is typically represented by several elements of $\boldsymbol{\beta}$. Suppose that we want to test

$$\text{H}_0 : \boldsymbol{\mu} = \mathbf{X}_0\boldsymbol{\beta}_0 \text{ against } \text{H}_1 : \boldsymbol{\mu} = \mathbf{X}\boldsymbol{\beta},$$

where $\mathbf{X}_0$ is 'nested' within $\mathbf{X}$ (meaning that $\mathbf{X}\boldsymbol{\beta}$ can exactly match any $\mathbf{X}_0\boldsymbol{\beta}_0$, but the reverse is not true). Without loss of generality we can assume that things are actually arranged so that $\mathbf{X} = [\mathbf{X}_0 : \mathbf{X}_1]$: it is always possible to re-parameterise the model so that this is the case. Suppose that $\mathbf{X}_0$ and $\mathbf{X}_1$ have $p - q$ and q columns, respectively, and let $\boldsymbol{\beta}_0$ and $\boldsymbol{\beta}_1$ be the corresponding subvectors of $\boldsymbol{\beta}$. The null hypothesis can hence be rewritten as $\text{H}_0 : \boldsymbol{\beta}_1 = \mathbf{0}$.

Now consider (7.2), the original QR decomposition of $\mathbf{X}$, in partitioned form:

$$\mathbf{X} = \mathbf{Q}\begin{bmatrix} \mathbf{R} \\ \mathbf{0} \end{bmatrix} \Rightarrow \mathbf{Q}^{\mathrm{T}}\mathbf{X} = \begin{bmatrix} \mathbf{R} \\ \mathbf{0} \end{bmatrix}$$
$$\Rightarrow \mathbf{Q}^{\mathrm{T}}[\mathbf{X}_0 : \mathbf{X}_1] = \begin{bmatrix} \tilde{\mathbf{R}}_0 : \mathbf{R}_1 \\ \mathbf{0} \end{bmatrix} \Rightarrow \mathbf{Q}^{\mathrm{T}}\mathbf{X}_0 = \begin{bmatrix} \tilde{\mathbf{R}}_0 \\ \mathbf{0} \end{bmatrix},$$

where $\tilde{\mathbf{R}}_0$ is the first $p - q$ columns of $\mathbf{R}$. Since $\mathbf{R}$ is upper triangular, the last q rows of $\tilde{\mathbf{R}}_0$ are 0, so let $\mathbf{R}_0$ denote the first $p - q$ rows of $\tilde{\mathbf{R}}_0$ (i.e. the first $p - q$ rows and columns of $\mathbf{R}$). Rotating $\mathbf{y} - \mathbf{X}_0\boldsymbol{\beta}_0$ using $\mathbf{Q}^{\mathrm{T}}$ implies that

$$\|\mathbf{y} - \mathbf{X}_0\boldsymbol{\beta}_0\|^2 = \left\|\mathbf{Q}^{\mathrm{T}}\mathbf{y} - \begin{bmatrix} \mathbf{R}_0 \\ \mathbf{0} \end{bmatrix}\boldsymbol{\beta}_0\right\|^2 = \|\mathbf{f}_0 - \mathbf{R}_0\boldsymbol{\beta}_0\|^2 + \|\mathbf{f}_1\|^2 + \|\mathbf{r}\|^2,$$

where $\mathbf{Q}^{\mathrm{T}}\mathbf{y}$ has been partitioned into $\mathbf{f}$ and $\mathbf{r}$, exactly as before, but $\mathbf{f}$ has then been further partitioned into $p - q$ vector $\mathbf{f}_0$ and q vector $\mathbf{f}_1$ so that $\mathbf{f} = \begin{bmatrix} \mathbf{f}_0 \\ \mathbf{f}_1 \end{bmatrix}$. Since the residual sum of squares for this null model is now $\|\mathbf{f}_1\|^2 + \|\mathbf{r}\|^2$, $\|\mathbf{f}_1\|^2$ is the increase in the residual sum of squares that results from dropping $\mathbf{X}_1$ from the model (i.e. from setting $\boldsymbol{\beta}_1 = \mathbf{0}$). That is, $\|\mathbf{f}_1\|^2$ is the difference in residual sum of squares between the 'full model' and the 'null model'.

Now, we know that $\mathbf{f} \sim N(\mathbf{R}\boldsymbol{\beta}, \mathbf{I}_p\sigma^2)$, but in addition we know that $\boldsymbol{\beta}_1 = \mathbf{0}$ under H_0 (i.e. the last q elements of $\boldsymbol{\beta}$ are zero). Hence

$$E\begin{bmatrix} \mathbf{f}_0 \\ \mathbf{f}_1 \end{bmatrix} = \mathbf{R}\boldsymbol{\beta} = (\tilde{\mathbf{R}}_0 : \mathbf{R}_1)\begin{bmatrix} \boldsymbol{\beta}_0 \\ \boldsymbol{\beta}_1 \end{bmatrix} = (\tilde{\mathbf{R}}_0 : \mathbf{R}_1)\begin{bmatrix} \boldsymbol{\beta}_0 \\ \mathbf{0} \end{bmatrix}$$
$$= \tilde{\mathbf{R}}_0\boldsymbol{\beta}_0 = \begin{bmatrix} \mathbf{R}_0 \\ \mathbf{0} \end{bmatrix}\boldsymbol{\beta}_0 = \begin{bmatrix} \mathbf{R}_0\boldsymbol{\beta}_0 \\ \mathbf{0} \end{bmatrix}.$$

So, if H_0 is true, $E(\mathbf{f}_1) = \mathbf{0}$ and $\mathbf{f}_1 \sim N(\mathbf{0}, \mathbf{I}_q\sigma^2)$. Consequently

$$\frac{1}{\sigma^2}\|\mathbf{f}_1\|^2 \sim \chi^2_q.$$

We also know that $\mathbf{f}_1$ and $\mathbf{r}$ are independent. So, forming an *F-ratio statistic*, assuming H_0 and using Section A.1.4, we have

$$F = \frac{\|\mathbf{f}_1\|^2/q}{\hat{\sigma}^2} = \frac{\frac{1}{\sigma^2}\|\mathbf{f}_1\|^2/q}{\frac{1}{\sigma^2}\|\mathbf{r}\|^2/(n-p)} \sim \frac{\chi^2_q/q}{\chi^2_{n-p}/(n-p)} \sim F_{q,n-p}, \quad (7.7)$$

and this result can be used to find the p-value for the hypothesis test. Remember that the term $\|\mathbf{f}_1\|^2$ is the difference in residual sum of squares between the two models being compared, and q is the difference in their degrees of freedom. So we could also write F as

$$F = \frac{(\|\mathbf{y} - \mathbf{X}_0\hat{\boldsymbol{\beta}}_0\|^2 - \|\mathbf{y} - \mathbf{X}\hat{\boldsymbol{\beta}}\|^2)/\{\dim(\boldsymbol{\beta}) - \dim(\boldsymbol{\beta}_0)\}}{\|\mathbf{y} - \mathbf{X}\hat{\boldsymbol{\beta}}\|^2/\{n - \dim(\boldsymbol{\beta})\}}.$$

7.1.5 The influence matrix

One useful matrix is the *influence matrix* (or *hat matrix*) of a linear model. This is the matrix that yields the fitted value vector, $\hat{\boldsymbol{\mu}}$, when post-multiplied by the data vector, $\mathbf{y}$. Recalling the definition of $\mathbf{Q}_f$, as being the first p columns of $\mathbf{Q}$, $\mathbf{f} = \mathbf{Q}_f^{\mathrm{T}}\mathbf{y}$, and so

$$\hat{\boldsymbol{\beta}} = \mathbf{R}^{-1}\mathbf{Q}_f^{\mathrm{T}}\mathbf{y}.$$

Furthermore $\hat{\boldsymbol{\mu}} = \mathbf{X}\hat{\boldsymbol{\beta}}$ and $\mathbf{X} = \mathbf{Q}_f\mathbf{R}$ so

$$\hat{\boldsymbol{\mu}} = \mathbf{Q}_f\mathbf{R}\mathbf{R}^{-1}\mathbf{Q}_f^{\mathrm{T}}\mathbf{y} = \mathbf{Q}_f\mathbf{Q}_f^{\mathrm{T}}\mathbf{y}.$$

So the matrix $\mathbf{A} \equiv \mathbf{Q}_f\mathbf{Q}_f^{\mathrm{T}}$ is the influence (hat) matrix such that $\hat{\boldsymbol{\mu}} = \mathbf{A}\mathbf{y}$.

The influence matrix has two interesting properties. First, the trace of the influence matrix is the number of (identifiable) parameters in the model, since

$$\text{tr}(\mathbf{A}) = \text{tr}(\mathbf{Q_f}\mathbf{Q_f}^{\text{T}}) = \text{tr}(\mathbf{Q_f}^{\text{T}}\mathbf{Q_f}) = \text{tr}(\mathbf{I_p}) = \mathbf{p}.$$

Second, $\mathbf{AA} = \mathbf{A}$, a property known as *idempotency*. Proof is simple:

$$\mathbf{AA} = \mathbf{Q}_f\mathbf{Q}_f^{\text{T}}\mathbf{Q}_f\mathbf{Q}_f^{\text{T}} = \mathbf{Q}_f\mathbf{I}_p\mathbf{Q}_f^{\text{T}} = \mathbf{Q}_f\mathbf{Q}_f^{\text{T}} = \mathbf{A}.$$

7.1.6 The residuals, $\hat{\epsilon}$, and fitted values, $\hat{\mu}$

The influence matrix is helpful in deriving properties of the fitted values, $\hat{\boldsymbol{\mu}}$, and the residuals, $\hat{\boldsymbol{\epsilon}}$. $\hat{\boldsymbol{\mu}}$ is unbiased, since $E(\hat{\boldsymbol{\mu}}) = E(\mathbf{X}\hat{\boldsymbol{\beta}}) = \mathbf{X}E(\hat{\boldsymbol{\beta}}) = \mathbf{X}\boldsymbol{\beta} = \boldsymbol{\mu}$. The covariance matrix of the fitted values is obtained from the fact that $\hat{\boldsymbol{\mu}}$ is a linear transformation of the random vector $\mathbf{y}$, which has covariance matrix $\mathbf{I}_n\sigma^2$, so that, using (1.5) from Section 1.5.1,

$$\mathbf{V}_{\hat{\mu}} = \mathbf{A}\mathbf{I}_n\mathbf{A}^{\text{T}}\sigma^2 = \mathbf{A}\sigma^2,$$

by the idempotence (and symmetry) of $\mathbf{A}$. The distribution of $\hat{\boldsymbol{\mu}}$ is degenerate multivariate normal.

Similar arguments apply to the residuals:

$$\hat{\boldsymbol{\epsilon}} = \mathbf{y} - \hat{\boldsymbol{\mu}} = (\mathbf{I} - \mathbf{A})\mathbf{y},$$

so

$$E(\hat{\boldsymbol{\epsilon}}) = E(\mathbf{y}) - E(\hat{\boldsymbol{\mu}}) = \boldsymbol{\mu} - \boldsymbol{\mu} = \mathbf{0}.$$

As in the fitted value case, we have

$$\mathbf{V}_{\hat{\epsilon}} = (\mathbf{I}_n - \mathbf{A})\mathbf{I}_n(\mathbf{I}_n - \mathbf{A})^{\text{T}}\sigma^2 = (\mathbf{I}_n - 2\mathbf{A} + \mathbf{AA})\,\sigma^2 = (\mathbf{I}_n - \mathbf{A})\,\sigma^2.$$

Again, the distribution of the residuals is degenerate normal. The results for the residuals are useful for model checking, because they allow the residuals to be standardised to have constant variance, if the model is correct.

7.1.7 The geometry of linear models

Least squares estimation of linear models amounts to finding the orthogonal projection of the n dimensional response data $\mathbf{y}$ onto the p dimensional linear subspace spanned by the columns of $\mathbf{X}$. The linear model states that $E(\mathbf{y})$ lies in the space spanned by all possible linear combinations of the columns of the $n \times p$ model matrix $\mathbf{X}$, and least squares seeks the point in

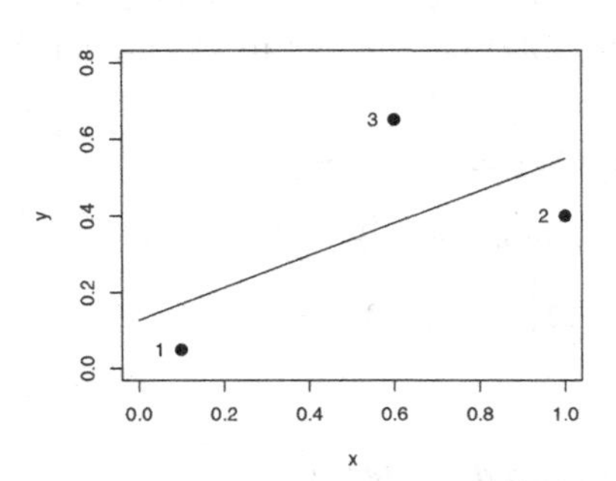

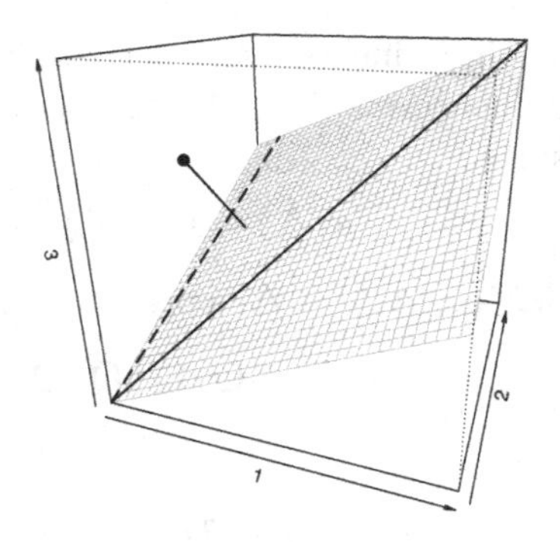

Figure 7.1 Illustration of the geometry of least squares. Left: a straight line fit to three x, y data. Right: the space in which the y coordinates of the data define a single point, while the columns of the model matrix (solid and dashed line) span the subspace shown in grey. The least squares estimate of $E(\mathbf{y})$ is the orthogonal projection of the data point onto the model subspace.

that space that is closest to $\mathbf{y}$ in Euclidean distance. Figure 7.1 illustrates this geometry for the model,

$$\begin{bmatrix} .05 \\ .40 \\ .65 \end{bmatrix} = \begin{bmatrix} 1 & .1 \\ 1 & 1 \\ 1 & .6 \end{bmatrix} \begin{bmatrix} \beta_0 \\ \beta_1 \end{bmatrix} + \begin{bmatrix} \epsilon_1 \\ \epsilon_2 \\ \epsilon_3 \end{bmatrix}.$$

7.1.8 Results in terms of $\mathbf{X}$

The presentation so far has been in terms of the method actually used to fit linear models in practice (employing the QR decomposition[5]). By taking this approach, results (7.6) and (7.7) can be derived concisely, without recourse to advanced linear algebra. However, for historical reasons, these results are more usually presented in terms of the model matrix, $\mathbf{X}$, rather than the components of its QR decomposition.

First consider the covariance matrix of $\hat{\boldsymbol{\beta}}$. This becomes $(\mathbf{X}^{\mathrm{T}}\mathbf{X})^{-1}\sigma^2$, which is easily seen to be equivalent to (7.4):

$$\begin{aligned} \mathbf{V}_{\hat{\beta}} &= (\mathbf{X}^{\mathrm{T}}\mathbf{X})^{-1}\sigma^2 = \left(\mathbf{R}^{\mathrm{T}}\mathbf{Q}_f^{\mathrm{T}}\mathbf{Q}_f\mathbf{R}\right)^{-1}\sigma^2 = \left(\mathbf{R}^{\mathrm{T}}\mathbf{R}\right)^{-1}\sigma^2 \\ &= \mathbf{R}^{-1}\mathbf{R}^{-\mathrm{T}}\sigma^2. \end{aligned}$$

[5] A few programs still fit models by solution of $\mathbf{X}^{\mathrm{T}}\mathbf{X}\hat{\boldsymbol{\beta}} = \mathbf{X}^{\mathrm{T}}\mathbf{y}$, but this is less computationally stable than the rotation method described here, although it is a bit faster.

The expression for the least squares estimates is $\hat{\beta} = (\mathbf{X}^{\mathrm{T}}\mathbf{X})^{-1}\mathbf{X}^{\mathrm{T}}\mathbf{y}$, which is equivalent to (7.3):

$$\hat{\beta} = (\mathbf{X}^{\mathrm{T}}\mathbf{X})^{-1}\mathbf{X}^{\mathrm{T}}\mathbf{y} = \mathbf{R}^{-1}\mathbf{R}^{-\mathrm{T}}\mathbf{R}^{\mathrm{T}}\mathbf{Q}_f^{\mathrm{T}}\mathbf{y} = \mathbf{R}^{-1}\mathbf{Q}_f^{\mathrm{T}}\mathbf{y} = \mathbf{R}^{-1}\mathbf{f}.$$

It follows that the influence matrix can be written as $\mathbf{A} = \mathbf{X}(\mathbf{X}^{\mathrm{T}}\mathbf{X})^{-1}\mathbf{X}^{\mathrm{T}}$. These results are of theoretical interest, but should not usually be used for computational purposes.

7.1.9 Interactions and identifiability

The preceding theory assumed that $\mathbf{X}$ has full rank. When working with factors some care is required to ensure that this happens. The issue is easiest to appreciate by considering the simple linear model

$$y_i = \alpha + \gamma_{k(i)} + \epsilon_i,$$

where α and the γ_k are parameters while $k(i)$ gives the group to which observation i belongs. Conceptually this model makes good sense: α is the overall mean, whereas γ_k is the departure from the overall mean caused by being a member of the kth group. The problem is that α and the γ_k are not identifiable. Any constant c could be added to α and simultaneously subtracted from all the γ_k, without changing the model-predicted distribution of the y_i. Hence there is no way that the model parameters can uniquely be determined from the data. This lack of identifiability leads directly to rank deficiency of $\mathbf{X}$, as is easily seen by writing out an example of the model matrix. Suppose, arbitrarily, that there are three groups, so that

$$\mathbf{X} = \begin{bmatrix} 1 & 1 & 0 & 0 \\ 1 & . & . & . \\ 1 & 1 & 0 & 0 \\ 1 & 0 & 1 & 0 \\ 1 & . & . & . \\ 1 & 0 & 1 & 0 \\ 1 & 0 & 0 & 1 \\ 1 & . & . & . \\ 1 & 0 & 0 & 1 \end{bmatrix}.$$

Any column of $\mathbf{X}$ is a simple linear combination of the other three, so that the matrix is rank 3. The lack of identifiability can be removed by placing a single linear constraint on the model parameters, the simplest of which is to set one of the parameters to zero. This could be α, but a choice that generalises to multiple factor models is to leave α free and to set the first level of the factor to zero, thereby removing the corresponding column of the model matrix and restoring full rank. If you write out an example model with m factor variables and an intercept, you will see that m constraints are required. Setting the first level of each factor to zero is a simple automatic way of generating them and is also the default in R. Notice that these constraints do not change what the model says about the distribution of the response. All that changes is the interpretation of the parameters: α is now the mean for the first level of the factor, whereas the γ_2, γ_3, etc. are the differences between each of the factor levels and the first.

Often in linear models we are interested in 'interaction' terms involving several predictors. Formally an interaction is generated in a model when the parameter for one predictor variable depends on another predictor variable (for example, the slope of a regression on age itself depends on the factor variable sex). It turns out that the model matrix columns associated with an interaction are given by all possible pairwise products of the model matrix columns for the effects that make up the interaction. Furthermore, if those effects are identifiable (perhaps by having had constraints imposed already), then the interactions constructed in this way are also identifiable. This is assuming that the data are sufficient to estimate the effect: for example, we cannot estimate the interaction coefficient associated with being over 50 and exercising for more than five hours a week from a sample that contains no individuals in this category.

As an example, consider a model with two factors and one metric variable, with an interaction of the factors and of the first factor with the metric variable. To save ink suppose that each factor has two levels. The model is

$$y_i = \alpha + \gamma_{k(i)} + \delta_{j(i)} + \eta_{k(i),j(i)} + \nu x_i + \omega_{k(i)} x_i + \epsilon_i.$$

Suppose there are 14 observations, the first 8 from the first level of the first factor, and the remainder from the second level, and that observations alternate between levels 1 and 2 of the second factor. Then the rank-deficient full model matrix is shown on the left, while a full rank version is shown on the right, using the simple constraints just described:

$$\begin{bmatrix}
1 & 1 & 0 & 1 & 0 & 1 & 0 & 0 & 0 & x_1 & x_1 & 0 \\
1 & 1 & 0 & 0 & 1 & 0 & 0 & 1 & 0 & x_2 & x_2 & 0 \\
1 & 1 & 0 & 1 & 0 & 1 & 0 & 0 & 0 & x_3 & x_3 & 0 \\
1 & 1 & 0 & 0 & 1 & 0 & 0 & 1 & 0 & x_4 & x_4 & 0 \\
1 & 1 & 0 & 1 & 0 & 1 & 0 & 0 & 0 & x_5 & x_5 & 0 \\
1 & 1 & 0 & 0 & 1 & 0 & 0 & 1 & 0 & x_6 & x_6 & 0 \\
1 & 1 & 0 & 1 & 0 & 1 & 0 & 0 & 0 & x_7 & x_7 & 0 \\
1 & 1 & 0 & 0 & 1 & 0 & 0 & 1 & 0 & x_8 & x_8 & 0 \\
1 & 0 & 1 & 1 & 0 & 0 & 1 & 0 & 0 & x_9 & 0 & x_9 \\
1 & 0 & 1 & 0 & 1 & 0 & 0 & 0 & 1 & x_{10} & 0 & x_{10} \\
1 & 0 & 1 & 1 & 0 & 0 & 1 & 0 & 0 & x_{11} & 0 & x_{11} \\
1 & 0 & 1 & 0 & 1 & 0 & 0 & 0 & 1 & x_{12} & 0 & x_{12} \\
1 & 0 & 1 & 1 & 0 & 0 & 1 & 0 & 0 & x_{13} & 0 & x_{13} \\
1 & 0 & 1 & 0 & 1 & 0 & 0 & 0 & 1 & x_{14} & 0 & x_{14}
\end{bmatrix}
\rightarrow
\begin{bmatrix}
1 & 0 & 0 & 0 & x_1 & 0 \\
1 & 0 & 1 & 0 & x_2 & 0 \\
1 & 0 & 0 & 0 & x_3 & 0 \\
1 & 0 & 1 & 0 & x_4 & 0 \\
1 & 0 & 0 & 0 & x_5 & 0 \\
1 & 0 & 1 & 0 & x_6 & 0 \\
1 & 0 & 0 & 0 & x_7 & 0 \\
1 & 0 & 1 & 0 & x_8 & 0 \\
1 & 1 & 0 & 0 & x_9 & x_9 \\
1 & 1 & 1 & 1 & x_{10} & x_{10} \\
1 & 1 & 0 & 0 & x_{11} & x_{11} \\
1 & 1 & 1 & 1 & x_{12} & x_{12} \\
1 & 1 & 0 & 0 & x_{13} & x_{13} \\
1 & 1 & 1 & 1 & x_{14} & x_{14}
\end{bmatrix}$$

Now consider a general $n \times p$ model matrix, $\mathbf{X}$, of rank $r < p$, with corresponding parameter vector $\boldsymbol{\beta}$. All that least squares does is to find the point in the space spanned by the columns of $\mathbf{X}$ that is as close as possible to $\mathbf{y}$ (in the Euclidean sense). So we could remove the rank-deficiency problem by defining $\boldsymbol{\beta} = \mathbf{C}\tilde{\boldsymbol{\beta}}$, where $\mathbf{C}$ is *any* $p \times r$ matrix such that $\tilde{\mathbf{X}} = \mathbf{XC}$ is of (full) rank r. $\tilde{\boldsymbol{\beta}}$ is the r vector of constrained parameters and $\tilde{\mathbf{X}}$ the corresponding model matrix. This observation implies that we have considerable freedom to define constraint matrices $\mathbf{C}$ so that the constrained parameters are interpretable. These alternative constrained parameterisations are known as alternative *contrasts*.

7.2 Linear models in R

The `lm` function in R is used to fit linear models to data and is the prototype for a large number of other functions for fitting standard classes of models. The first argument to `lm` is a *model formula* that specifies the response variable and the structure of the model matrix. The second, optional, argument is a dataframe containing the variables referred to by the model formula. `lm` estimates the model using exactly the QR method covered earlier, having first imposed any necessary identifiability constraints. It returns a fitted model object of class `"lm"`.

The returned fitted model object can be interrogated by various method functions for printing, summarizing, producing residual plots and so on. Here is a short example in which data are simulated from

$$y_i = \alpha + \gamma_{k(i)} + \delta x_i + \epsilon_i,$$

and the parameters are then estimated from the resulting data by least squares:

```
> set.seed(0);g <- rep(1:3,10); x <- runif(30)
> y <- 1 + x + g + rnorm(30) * 0.5
> g <- factor(g)
> dat <- data.frame(y=y,g=g,x=x)
>
> mod <- lm(y ~ g + x,dat)
> mod ## causes print method to be called for mod

Call:
lm(formula = y ~ g + x, data = dat)

Coefficients:
(Intercept)           g2           g3            x
     2.0362       0.9812       2.1461       0.8590
```

The structure of the model is specified using the formula `y ~ g + x`, where the fact that `g` is declared to be of class `"factor"` causes it to be treated as such in the model fitting. `lm` has automatically implemented identifiability constraints here, setting the first coefficient for factor `g` to zero, so that the intercept is now the intercept for level 1 of factor `g`.

A more extensive summary of the model can also be obtained as follows:

```
> summary(mod)

Call:
lm(formula = y ~ g + x, data = dat)

Residuals:
    Min       1Q   Median       3Q      Max
-0.64293 -0.26466 -0.07511  0.27505 0.89931

Coefficients:
          Estimate Std. Error t value Pr(>|t|)
(Intercept) 2.0362  0.2023  10.067  1.84e-10 ***
g2          0.9812   0.1873  5.237  1.80e-05 ***
g3          2.1461   0.1849  11.605  8.76e-12 ***
x           0.8590   0.2579  3.331  0.0026 **
--
Signif. codes: 0 '***' 0.001 '**' 0.01 '*' 0.05 '.' 0.1

Residual standard error: 0.411 on 26 degrees of freedom
Multiple R-squared: 0.8436, Adjusted R-squared: 0.8256
F-statistic: 46.75 on 3 and 26 DF, p-value: 1.294e-10
```

After printing the model call and a summary of the residual distribution, the coefficient table gives the parameter estimates and their standard errors, as well as a t statistic for testing each parameter for equality to zero and the p-value for such a test (based on Section 7.1.3). Notice how coefficients are identified by the name of the predictor variable with which they are associated (R knows nothing about what we might choose to call the coefficients; it only knows about the associated variables). The output from this table can also be used to compute confidence intervals for the model coefficients using the results from Section 2.7. The `Residual standard error` is $\hat{\sigma}$, and the `F-statistic` is the F-ratio for testing the null hypothesis that a constant is as good a model for the mean response as the model actually fitted. The associated p-value suggests not, in this case.

The `R-squared` statistics are measures of how closely the model fits the response data. The idea is that, after fitting, the part of the variability left unexplained is the variability of the residuals, so the proportion of variability unexplained is the ratio of the residual variance to the original variance of the y_i. One minus the unexplained variance is the explained variance

$$r^2 = 1 - \frac{\sum_i \hat{\epsilon}_i^2/n}{\sum_i (y_i - \bar{y})^2/n}.$$

This conventional definition (from which the n's can be cancelled) uses biased variance estimators. As a result r^2 tends to overestimate how well a model is doing. The *adjusted* r^2 avoids this overestimation to some extent by using unbiased estimators,

$$r^2_{\text{adj}} = 1 - \frac{\sum_i \hat{\epsilon}_i^2/(n-p)}{\sum_i (y_i - \bar{y})^2/(n-1)},$$

where p is the number of model parameters. r^2_{adj} can be negative.[6]

High r^2 values (close to 1) indicate a close fit, but a low r^2 is not necessarily indicative of a poor model: it can simply mean that the data contain a substantial random component.

7.2.1 Model formulae

In R, model formulae are used to specify the response variable and model structure. Consider the example

```
y ~ x + log(z) + x:z
```

The variable to the left of `~` specifies the response variable, whereas everything to the right specifies how to set up the model matrix. '`+`' indicates to include the variable to the left of it *and* the variable to the right of it (it does not mean that they should be summed). '`:`' denotes the interaction of the variables to its left and right. So if x is a metric variable then the above formula specifies:

$$y_i = \beta_1 + \beta_2 x_i + \beta_3 \log(z_i) + \beta_4 x_i z_i + \epsilon_i$$

whereas if x is a factor variable the model is

$$y_i = \beta_1 + \gamma_{k(i)} + \beta_2 \log(z_i) + \alpha_{k(i)} z_i + \epsilon_i.$$

Notice how an intercept term is included by default.

In addition to '`+`' and '`:`' several other symbols have special meanings in model formulae:

- '`*`' means to include main effects and interactions, so `a * b` is the same as `a + b + a:b`.

[6] This occurs when the fit of the model to data is purely imaginary.

- '^' is used to include main effects and interactions up to a specified level. So `(a+b+c)^2` is equivalent to `a + b + c + a:b + a:c + b:c`, for example, while `(a+b+c)^3` would also add `a:b:c`. Notice that this operator does not generate all the second-order terms you might be expecting for metric variables.
- '-' excludes terms that might otherwise be included. For example, `-1` excludes the intercept otherwise included by default, and `x * z - z` would produce `x + x:z`.

As we have seen, you can use simple mathematical functions in model formulae to transform variables, but outside the argument to a function the usual arithmetic operators all have special meanings. This means that if we want to restore the usual arithmetic meaning to an operator in a formula, then we have to take special measures to do this, by making the expression the argument of the identity function `I()`. For example, `y ~I(x+z)` would specify the model $y_i = \alpha + \beta(x_i + z_i) + \epsilon_i$. Occasionally the model matrix should include a column for which the corresponding β coefficient is fixed at 1. Such a column is known as an *offset*: `offset(z)` would include a column, `z`, of this type. See `?formula` in R for more details.

7.2.2 Model checking

As with all statistical modelling it is important to check the plausibility of a linear model before conducting formal statistical inference. It is pointless computing AIC, testing hypotheses or obtaining confidence intervals for a model that is clearly wrong, because all these procedures rest on the model being at least plausible. For linear models the key assumptions are those of constant variance and independence, and the residuals should be examined for any evidence that these have been violated. Normality of the residuals should also be checked, if the other assumptions are viable, but the central limit theorem tends to mean that normality is of only secondary importance to the other assumptions.

Often the constant variance assumption is violated because the variance actually depends on the mean of the response, so plots of $\hat{\epsilon}_i$ against $\hat{\mu}_i$, can be very useful. Independence tends to be violated when observations that are nearby in space or time are correlated, or when something is wrong with the mean structure of the model, such as a predictor having been omitted, or included incorrectly (specified as a linear effect when a quadratic was appropriate, for example). Plots of residuals against predictors and potential predictors are useful, as well as estimation of the degree of correlation in space and time.

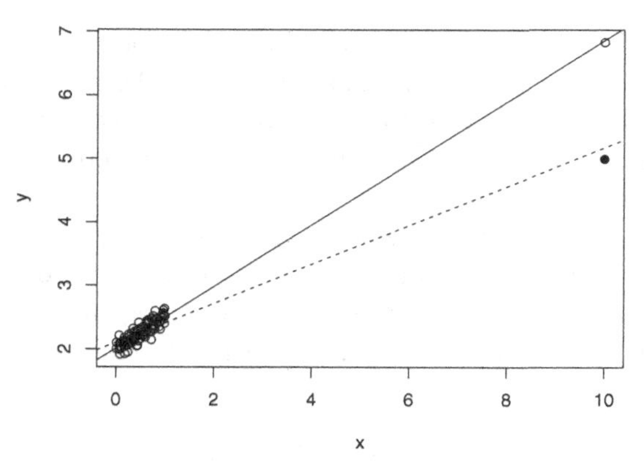

Figure 7.2 Leverage: the basic problem. The solid black line shows the least squares straight line fit to the 100 data shown as open circles. In this case the fit is reasonable for all the data. The dashed line shows the least squares straight line fit when the point at $x = 10$ is moved to the position shown as a solid black disc. To accommodate the datum far from the other points, the line has been forced to fit the remaining 99 data rather poorly.

Another thing to check is whether some individual observations are having undue influence on the modelling results. Points with very large residuals can sometimes be problematic: perhaps they are recorded incorrectly or are observations that simply do not belong in the same population as the rest of the data. Large outliers should be investigated. Sometimes there is something wrong with the corresponding observations, justifying exclusion of these points from the analysis, but sometimes, on closer investigation, these are the observations that contain the most interesting information in the dataset. If there appears to be nothing unusual about outliers apart from them being outliers, then it is prudent to repeat the analysis with and without them, to check the sensitivity of conclusions to these points. However, outliers should almost never be discarded simply for being outliers (if Geiger and Marsden had discarded the outliers in Rutherford's 1909 experiment, they would have left the nucleus of the atom undiscovered).

A related issue is that of leverage: some points have undue influence not because their response variable is noticeably out of line, but because an unusual combination of predictor variables makes the whole fit unduly sensitive to the value of the corresponding response observation. Figure 7.2 illustrates this issue.

As an example of some basic model checking, consider a model for the `cars` data supplied with R. The data give stopping `dist`ance in feet for cars stopping from a range of `speed`s in miles per hour and were gathered in the 1920s. Theoretically the stopping distance for a car is made up of a driver reaction distance and a braking distance. The former comes from the fixed length of time taken for the driver to respond to the stop signal and apply the brakes, so this distance should be directly proportional to speed. Once the brakes are applied, the distance is determined by the rate at which the brakes can dissipate the car's kinetic energy. Brakes dissipate kinetic energy in direct proportion to the distance travelled, and the total amount of energy to dissipate is proportional to the square of speed, so this component of the distance should be proportional to initial speed squared, suggesting a model

$$\texttt{dist}_i = \beta_0 + \beta_1 \texttt{speed}_i + \beta_2 \texttt{speed}_i^2 + \epsilon_i$$

(if the reasoning behind the model is correct then β_0 should be approximately zero). Let us fit this model to the `cars` data and then examine the default residual plots for an `"lm"` object:

```
b <- lm(dist ~ speed + I(speed^2),data=cars)
par(mfrow=c(2,2))
plot(b)
```

Figure 7.3 shows the results. The top left plot of $\hat{\epsilon}_i$ against $\hat{\mu}_i$ shows some indication of increasing variance with mean, which would somewhat violate the constant variance assumption, although the effect is not extreme here. The other feature to look for is a pattern in the average value of the residuals as the fitted values change. The solid curve shows a running average of the residuals to help judging this: there is no obvious pattern here, which is good. The remaining plots shows standardised residuals $\hat{\epsilon}_i/(\hat{\sigma}\sqrt{1 - A_{ii}})$, which should appear approximately $N(0, 1)$ distributed if the model is correct (see Section 7.1.6). The lower left plot shows the square root of the absolute value of the standardised residuals against the fitted value (again with a running average curve). If all is well, the points should be evenly spread with respect the vertical axis here, with no trend in their average value. A trend in average value is indicative of a problem with the constant variance assumption, and is clearly visible in this case. The top right plot shows the ordered standardised residuals against quantiles of a standard normal: the systematic deviation from a straight line at the top right of the plot indicates a departure from normality in the residuals. The lower right plot is looking at leverage and influence of residuals, by plotting standardised residuals against a measure of leverage, A_{ii}. A combination of high residuals and high leverage indicates a point with substantial influence on the fit. A standard way of measuring this is via *Cook's distance*, which measures the change in all model fitted-values on omission

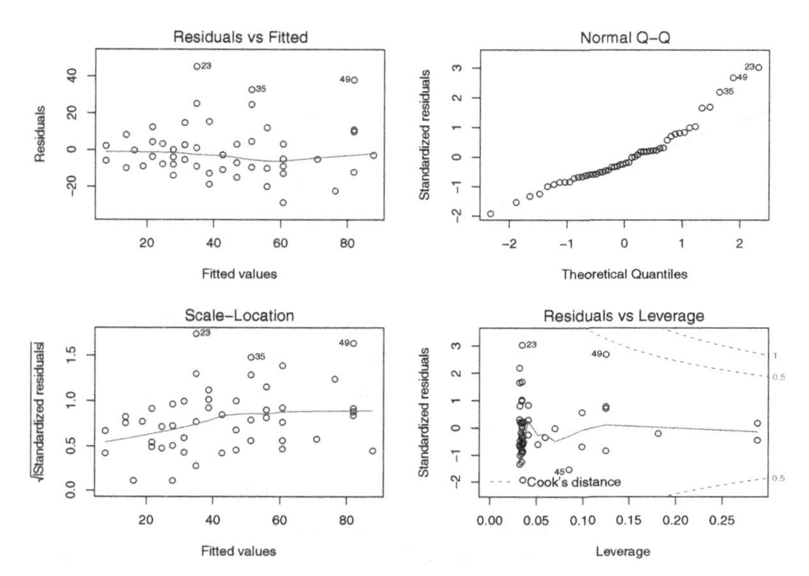

Figure 7.3 Default model checking plots for the `cars` model discussed in Section 7.2.2. The left-hand plots both suggest increasing variance with the mean and the top right plot indicates some departure from normality in the residuals. The lower right plot suggests that, although a couple of points have rather high leverage, their actual influence on the fit is not unduly large.

of the data point in question. It turns out that Cook's distance is a function of A_{ii} and the standardised residuals, so contours of Cook's distance values are shown on the plot. Cook's distances over 0.5 are considered borderline problematic, whereas values over 1 are usually considered highly influential, so points to the right of these contours warrant investigation. Here there seems to be no problem.

Given these plots, an obvious model to try is one in which variability increases with speed; for example, $\epsilon_i \sim N(0, \sigma^2 \texttt{speed}_i)$.

```
lm(dist ~ speed + I(speed^2),data=cars,weights=1/speed)
```

would fit this and does indeed improve matters, but it is time to move on, noting that in most serious analyses we would need to plot residuals against predictors, rather than relying solely on the default plots.

7.2.3 Prediction

After fitting a model, one common task is to predict the expected response from the model at new values of the predictor variables. This is

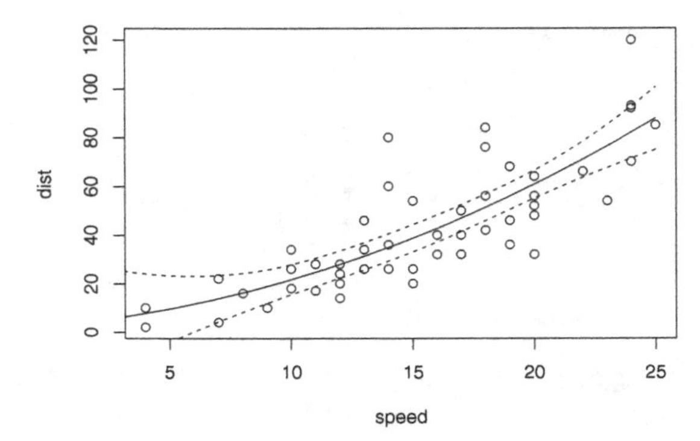

Figure 7.4 Fitted car model predictions overlaid on observed data, as discussed in Section 7.2.3.

easy: simply use the new predictor variable values to create a *prediction matrix*, $\mathbf{X}^p$, in exactly the same way as the original values were used to create $\mathbf{X}$. Then the predictions are $\hat{\boldsymbol{\mu}}^p = \mathbf{X}^p\hat{\boldsymbol{\beta}}$, and $\hat{\boldsymbol{\mu}}^p \sim N(\boldsymbol{\mu}^p, \mathbf{X}^p(\mathbf{X}^\mathrm{T}\mathbf{X})^{-1}\mathbf{X}^{pT}\sigma^2)$. In R the method function `predict.lm` automates the process. The following code uses it to add a predicted distance curve, with 2 standard error bands, to a plot of the cars data:

```
with(cars,plot(speed,dist))
dat <- data.frame(speed = seq(0,25,length=100))
fv <- predict(b,newdata=dat,se=TRUE)
lines(dat$speed,fv$fit)
lines(dat$speed,fv$fit + 2 * fv$se.fit,lty=2)
lines(dat$speed,fv$fit - 2 * fv$se.fit,lty=2)
```

The results are shown in Figure 7.4.

7.2.4 Interpretation, correlation and confounding

On examination of the summary of the `cars` model, something strange stands out:

```
> summary(b)

Call:
lm(formula = dist ~ speed + I(speed^2), data = cars)

Residuals:
   Min      1Q Median     3Q    Max
-28.720 -9.184 -3.188 4.628 45.152

Coefficients:
            Estimate Std. Error t value Pr(>|t|)
```

```
(Intercept) 2.47014 14.81716 0.167 0.868
speed      0.91329 2.03422 0.449 0.656
I(speed^2) 0.09996 0.06597 1.515 0.136

Residual standard error: 15.18 on 47 degrees of freedom
Multiple R-squared: 0.6673, Adjusted R-squared: 0.6532
F-statistic: 47.14 on 2 and 47 DF, p-value: 5.852e-12
```

The p-values for all the model terms are very high, despite the fact that the predictions from the model as a whole clearly indicate that there is good evidence that the model is better than zero or a constant model. These p-values cannot be taken as an indication that all the terms can be dropped from the model, but why not? The answer is that the p-values are testing whether the corresponding coefficients could really be zero *given that the other terms remain in the model* (i.e. are nonzero). If the estimators for the various coefficients are not independent, then dropping one term (setting it to zero) will change the estimates of the other coefficients and hence their p-values. For this reason, if we were to consider dropping terms then we should drop only one at a time, refitting after each drop. It often makes sense for the single dropped term to be the one with the highest p-value. Only if all the coefficient estimators are independent can we dispense with this cautious approach and drop all the terms with high p-values from a model in one go. However, such independence usually only arises for models of 'balanced' data from experiments designed to achieve it.

Lack of independence between estimators creates difficulties in the interpretation of estimates. The basic issue is that correlation between parameter estimators typically arises from correlation between the variables to which the parameters relate, but if predictor variables are correlated it is not possible to entirely separate out their effects on the response by examining the results of model fitting. As an example, consider modelling blood pressure in a group of patients using the predictor variables height and weight. Here is a simple simulation in which the real driver of blood pressure is weight, but height and weight are correlated:

```
n <- 50; set.seed(7)
height <- rnorm(n,180,10)
weight <- height^2/400+rnorm(n)*5
bp <- 80 + weight/2 + rnorm(n)*10
```

Now fit $\texttt{bp}_i = \beta_0 + \beta_1\texttt{height}_i + \beta_2\texttt{weight}_i + \epsilon_i$, and summarise it:

```
> summary(lm(bp~height+weight))
...
Coefficients:
            Estimate Std. Error t value Pr(>|t|)
```

```
(Intercept) 32.76340 30.57422 1.072 0.2894
height    0.45497 0.26894 1.692 0.0973 .
weight    0.09462 0.27248 0.347 0.7299
...
```

In this case most of the effect on blood pressure has been attributed to `height`, basically because the correlation between height and weight is about 0.9, so that it is not possible to determine which variable is actually driving the response. Comparison with the two single-effect models emphasises the difficulty:

```
> summary(lm(bp~height))
...
Coefficients:
         Estimate Std. Error t value Pr(>|t|)
(Intercept) 25.4937 22.0782 1.155 0.254
height     0.5382   0.1209 4.453 5.05e-05 ***
...
> summary(lm(bp~weight))
...
Coefficients:
         Estimate Std. Error t value Pr(>|t|)
(Intercept) 81.3987 10.6049 7.676 6.83e-10 ***
weight     0.5054   0.1260 4.012 0.00021 ***
...
```

Notice how both model coefficients were modified by the presence of the other correlated predictor (the true coefficient values being 0 for `height` and 0.5 for `weight`). Another way of looking at this is that `height` is such a good proxy for `weight` in these data that we can't tell whether weight or its proxy is the better predictor. On reflection, the only way to determine which is the causal variable here is to control for the other one. That is, to find a way of comparing different heights with weight held constant, and different weights with height held constant. Case control studies operate this way, by trying to identify pairs of patients matched on every relevant variable except for the variable being investigated.

The obvious related problem is the problem of hidden confounders, which are variables not in the model that are related to both the response and one or more of the predictors that are included. Since the included predictors act as proxies for the confounders, their coefficient estimates are distorted by including both a component relating to their direct affect, and a component relating to their effect as a proxy for the confounder.

The issue of hidden confounding and correlation is a major reason for basing causal inference (such as deciding 'does this drug work or not?')

on designed experiments. With appropriate design we can ensure that parameters associated with the different effects controlled for in the experiment are independent. Furthermore, by random allocation of experimental units (e.g. patients) to the different levels of the factors controlled for, we can break any association between the factor variables controlled by the experiment and variables that could otherwise be confounders. See Section 2.6.

7.2.5 Model comparison and selection

The results of Section 7.1.4 allow the comparison of nested linear models by hypothesis testing, and the R function `anova` automates this. As an example consider testing the null model $\texttt{dist}_i = \beta\texttt{speed}_i^2 + \epsilon_i$ against the full model for the `cars` data considered earlier. The following performs the appropriate F-ratio test (using the variance modification suggested by the model checking):

```
> b <- lm(dist~speed+I(speed^2),data=cars,weights=1/speed)
> b0 <- lm(dist~I(speed^2)-1 ,data=cars,weights=1/speed)
> anova(b0,b)
Analysis of Variance Table

Model 1: dist ~ I(speed^2) - 1
Model 2: dist ~ speed + I(speed^2)
  Res.Df RSS Df Sum of Sq F Pr(>F)
1     49 756.11
2     47 663.42 2 92.693 3.2834 0.04626 *
```

so there is some evidence against the null model in this case. If `anova` is called with a single argument, then a table is produced based on the sequence of ever simpler models obtained by removing terms sequentially from the model. Each row of the table tests one model in the sequence against the closest more complicated model in the sequence. Such tables only really make sense in the context of balanced designed experiments where the effects are independent. Otherwise the `drop1` function is usually a better bet: it produces the table obtained by F-ratio test comparison of the full model with each of the models produced by dropping a single effect from the full model.

The `AIC` function compares models by AIC (see Section 4.6). For example,

```
> AIC(b0,b)
   df      AIC
b0  2 414.8026
b   4 412.2635
```

which again suggests that that the larger model is preferable here. `BIC` is also available.

Model selection strategies

When faced with large numbers of possible prediction terms in a model, model comparison methods are often used to try and sort through the space of possible models to find one that is 'best' in some sense. A traditional approach is *backwards* selection, which starts with the 'largest plausible model' and consists of repeatedly deleting the model term with the highest p-value (as reported by `drop1`) and refitting, until all p-values are below some threshold. *Forward selection* starts from a simple model and repeatedly adds in the single predictor term for which there is most evidence in an F-ratio test, until no more terms would lead to significant improvement. Forward selection is slightly problematic theoretically, because early in the process it is likely that both models being compared are demonstrably wrong, which invalidates the theoretical basis for the test. Perhaps more seriously, early in the process the residual variance may be seriously inflated as a result of important terms not yet being included in the model, which means that the early tests lack power and termination may occur far too soon. Pragmatically, however, it may be the only solution for large problems. Naturally there are also *backward-forward* strategies, in which cycles of backward and forward selection are alternated until convergence, to allow terms that were dropped early on the possibility of re-entering the model.

Selection strategies based on hypothesis testing are somehow searching for the simplest model compatible with the data. As an alternative to this approach we can also use AIC to compare the alternative models. The `step` function in R automates the process of backward, forward, or backward-forward selection based on AIC. In simple cases it may also be possible to fit all possible submodels of some initial largest model, and just select the submodel with the smallest AIC.

Another selection approach, which has gained popularity when there are large numbers of predictors relative to the number of data, is to penalise the model coefficients towards zero, in such a way that as the penalization increases, many of the coefficient estimates become zero (see e.g. Hastie et al., 2001). For example, the model fitting problem becomes

$$\hat{\boldsymbol{\beta}} = \underset{\beta}{\operatorname{argmin}} \|\mathbf{y} - \mathbf{X}\beta\|^2 + \lambda \sum_{i=1}^{p} |\beta_i|,$$

where penalisation parameter λ is increased to force successively more terms out of the model. Obviously, care must be taken to standardise the predictors appropriately for this *Lasso* method to make sense.

7.3 Extensions

Linear models have proved so useful that they have been generalised in several ways.

- *Linear mixed models* augment the linear model structure with a much richer linear structure for the random variability in the data (e.g. Pinheiro and Bates, 2000). The basic model becomes

$$\mathbf{y} = \mathbf{X}\boldsymbol{\beta} + \mathbf{Z}\mathbf{b} + \boldsymbol{\epsilon}, \quad \mathbf{b} \sim N(\mathbf{0}, \boldsymbol{\psi}), \quad \boldsymbol{\epsilon} \sim N(\mathbf{0}, \mathbf{I}\sigma^2),$$

  where $\boldsymbol{\beta}$, σ^2 and $\boldsymbol{\psi}$ are now parameters ($\boldsymbol{\psi}$ usually has some structure so that it actually depends on only a small set of parameters $\boldsymbol{\theta}$). $\mathbf{Z}$ is a model matrix specifying how the stochastic structure of the response depends on the random effects, $\boldsymbol{b}$. Inference is now based on maximum likelihood estimation using the fact that $\mathbf{y} \sim N(\mathbf{X}\boldsymbol{\beta}, \mathbf{I}\sigma^2 + \mathbf{Z}\boldsymbol{\psi}\mathbf{Z}^{\mathrm{T}})$ (and in practice usually exploiting any special structure in $\mathbf{Z}$ and $\boldsymbol{\psi}$). One interesting and computationally useful fact is that, given the other parameters, $\hat{\boldsymbol{\beta}}$ and the modes of $\mathbf{b}|\mathbf{y}$ are the minimisers of

$$\|\mathbf{y} - \mathbf{X}\boldsymbol{\beta} - \mathbf{Z}\mathbf{b}\|^2/\sigma^2 + \mathbf{b}^{\mathrm{T}}\boldsymbol{\psi}^{-1}\mathbf{b},$$

  a penalized least squares problem. See `lme` from library `nlme` in R.
- *Generalised linear models* (GLMs) allow the response variable to have any exponential family distribution (Poisson, gamma, binomial, etc.) while some nonlinearity is allowed into the mean structure (McCullagh and Nelder, 1989). Defining $\mu_i = E(y_i)$ a generalised linear model has the form

$$g(\mu_i) = \mathbf{X}_i\boldsymbol{\beta}, \quad y_i \sim \mathrm{EF}(\mu_i, \phi),$$

  where g is some known monotonic function (identity or log, for example), $\mathbf{X}_i$ is the ith row of $\mathbf{X}$, and $\mathrm{EF}(\mu_i, \phi)$ denotes some exponential family distribution with mean μ_i and scale parameter ϕ. $\mathbf{X}\boldsymbol{\beta}$ is known as the *linear predictor* for such models and is often denoted $\boldsymbol{\eta}$. Maximum likelihood estimation theory provides the basis for model estimation and further inference for these models, but there are many links to the linear models of this chapter. In particular, noting that for any exponential family distribution $\mathrm{var}(y_i) = V(\mu_i)\phi$, where V is a known function, it turns out that MLE by Newton's method is equivalent to the iterative estimation of working weighted linear models, as follows (here using the expected Hessian). Set $\hat{\mu}_i = y_i + \Delta_i$ and $\hat{\eta}_i = g(\hat{\mu}_i)$ (where Δ_i is a small perturbation that may be made to ensure the existence of $\hat{\eta}_i$) and iterate the following two steps to convergence:

1. For $i = 1, \ldots, n$ form $\hat{\eta}_i = g(\hat{\mu}_i)$, $z_i = g'(\hat{\mu}_i)(y_i - \hat{\mu}_i) + \hat{\eta}_i$, and $w_i = V(\hat{\mu}_i)^{-1} g'(\hat{\mu}_i)^{-2}$.
2. Compute

$$\hat{\boldsymbol{\beta}} = \underset{\boldsymbol{\beta}}{\text{argmin}} \sum_{i=1}^{n} w_i (z_i - \mathbf{X}_i \boldsymbol{\beta})^2$$

and update $\hat{\eta}_i = \mathbf{X}_i \hat{\boldsymbol{\beta}}$, and $\hat{\mu}_i = g^{-1}(\hat{\eta}_i)$.

At convergence (4.5) becomes $\hat{\boldsymbol{\beta}} \sim N(\boldsymbol{\beta}, (\mathbf{X}^{\mathrm{T}} \mathbf{W} \mathbf{X})^{-1} \phi)$, where $\mathbf{W}$ is the diagonal matrix of converged w_i (this is a large sample approximation, of course). See `glm` in R for fitting these models.

- *Generalised additive models* (GAMs) are generalised linear models in which the linear predictor depends linearly on unknown smooth functions of predictors (e.g. Wood, 2006). In general the model becomes

$$g(\mu_i) = \mathbf{X}_i^* \boldsymbol{\beta}^* + \sum_j L_{ij} f_j, \quad y_i \sim \text{EF}(\mu_i, \phi)$$

where $\mathbf{X}_i^* \boldsymbol{\beta}^*$ is the parametric component of the linear predictor (often just an intercept term), the f_k are smooth functions of one or more predictors, and the L_{ij} are linear functionals. The most common example is where the L_{ij} are functionals of evaluation, so that the model is

$$g(\mu_i) = \mathbf{X}_i^* \boldsymbol{\beta}^* + \sum_j f_j(x_{ji}), \quad y_i \sim \text{EF}(\mu_i, \phi)$$

where x_{ji} denotes the ith observation of the jth predictor (possibly a vector). The model is accompanied by precise characterisation of what is meant by a *smooth* function, in the form of measures of function wiggliness such as the spline penalty $\int f''(x)^2 dx$.

The models offer convenient flexibility relative to purely parametric GLMs, but at the cost that the functions must be estimated, which includes estimating how smooth they should be. A convenient approach replaces each f_j with a linear basis expansion $f_j(x) = \sum_{k=1}^{K} b_{jk}(x) \gamma_k$ where the γ_k are coefficients to be estimated, and the $b_{jk}(x)$ are basis functions, chosen to have good approximation theoretic properties (such as spline bases). K is chosen to strike a balance between avoiding approximation error bias and achieving computational efficiency. Given the basis expansion, the GAM now has the form of a rather richly parameterised GLM, $g(\mu_i) = \mathbf{X}_i\boldsymbol{\beta}$, where $\mathbf{X}$ contains the original $\mathbf{X}^*$ and columns containing each of the basis functions evaluated at the covariate values (or linear functionals of these) and $\boldsymbol{\beta}$ contains the collected parameters. The penalties become quadratic forms, $\boldsymbol{\beta}^{\mathrm{T}}\mathbf{S}_j\boldsymbol{\beta}$, where the $\mathbf{S}_j$ are matrices of known coefficients.

To avoid overfit, estimation is by penalised maximum likelihood estimation, so we seek

$$\hat{\boldsymbol{\beta}} = \underset{\beta}{\operatorname{argmax}}\, l(\boldsymbol{\beta}) - \frac{1}{2}\sum_j \lambda_j \boldsymbol{\beta}^{\mathrm{T}}\mathbf{S}_j\boldsymbol{\beta}.$$

The λ_j are tunable *smoothing parameters* controlling the fit-smoothness tradeoff. In fact, given values for the smoothing parameters, $\hat{\boldsymbol{\beta}}$ can be obtained by a penalised version of the iterative weighted least squares method used to fit GLMs. All that changes is that at the second step of the algorithm we have

$$\hat{\boldsymbol{\beta}} = \underset{\beta}{\operatorname{argmin}} \sum_{i=1}^{n} w_i(z_i - \mathbf{X}_i\boldsymbol{\beta})^2 + \sum_j \lambda_j \boldsymbol{\beta}^{\mathrm{T}}\mathbf{S}_j\boldsymbol{\beta}.$$

To estimate the λ_j there are two main approaches. The first chooses the λ_i to optimise an estimate of how well the model would fit new data (not used in the estimation), such as AIC or some cross-validation score. The alternative treats the penalties as being induced by improper Gaussian priors on the model coefficients $\boldsymbol{\beta} \sim N\{\mathbf{0}, \tau(\sum_j \lambda_j \mathbf{S}_j)^{-}\}$ (the covariance matrix here is a pseudoinverse of the total penalty matrix). A Laplace approximation can then be used to integrate the $\boldsymbol{\beta}$ out of the marginal likelihood for the λ_j and ϕ, and this marginal likelihood can be maximised to estimate the scale parameter (if needed) and the smoothing parameters. The computations resemble those used for estimating the variance parameters of random effects, but some care is needed in the interpretation. It is rarely the case that the modeller believes that the f_k would be resampled afresh from their priors on each replication of the data, so the procedure is really best thought of as Bayesian. In any case the f_k have the dual interpretation of being smooth functions and posterior modes of random fields. Inference is most usefully based on the large-sample Bayesian result $\boldsymbol{\beta} \sim N\{\hat{\boldsymbol{\beta}}, (\mathbf{X}^{\mathrm{T}}\mathbf{W}\mathbf{X} + \sum_j \lambda_j \mathbf{S}_j)^{-1}\phi\}$. See function `gam` in R package `mgcv`.

Unsurprisingly, these various extensions have been combined, yielding generalised linear mixed models (GLMMs) and generalised additive mixed models (GAMMs), for example. Given the link between estimating smooth functions and estimating random effects, these latter extensions are computable by methods almost identical to those used for GAMs. Full Bayesian approaches to these model classes are also available via MCMC (see e.g. Fahrmeir et al., 2004, the stand alone BayesX package, and its R interface), or higher order approximation based on nested Laplace approximation (see Rue et al., 2009, the stand alone INLA package and its R interface).

Exercises

7.1 Find an expression for the least squares estimate of β in the model $y_i = \beta x_i = \epsilon_i$, in terms of x_i and y_i, by minimising $\sum_i (y_i - \beta x_i)^2$ w.r.t. β.

7.2 This question provides an alternative derivation of the least squares estimates. Let $\mathcal{S}(\boldsymbol{\beta})$ denote $\|\mathbf{y} - \mathbf{X}\boldsymbol{\beta}\|^2$. If $\mathbf{X}^{\mathrm{T}}\mathbf{X}\boldsymbol{\beta}_0 = \mathbf{X}^{\mathrm{T}}\mathbf{y}$, show that

$$\mathcal{S}(\boldsymbol{\beta}) - \mathcal{S}(\boldsymbol{\beta}_0) = \|\mathbf{X}(\boldsymbol{\beta} - \boldsymbol{\beta}_0)\|^2.$$

What does this tell you about $\boldsymbol{\beta}_0$?

7.3 Show that

$$\hat{\sigma}^2 = \frac{\|\mathbf{r}\|^2}{n - p}$$

is an unbiased estimator of the residual variance σ^2 by considering only $E(r_i^2)$ (and not assuming normality).

7.4 Show that, in the usual linear modelling notation, $\mathbf{X}^{\mathrm{T}}\mathbf{y} = \mathbf{X}^{\mathrm{T}}\hat{\boldsymbol{\mu}}$. Hence if a linear model contains an intercept term, show what the sum of the residuals, $\sum_i \hat{\epsilon}_i$, must be.

7.5 Write out the following three models in the form $\mathbf{y} = \mathbf{X}\boldsymbol{\beta} + \boldsymbol{\epsilon}$ (note: $\mathbf{y}$, $\boldsymbol{\beta}$ and $\boldsymbol{\epsilon}$ are always vectors, whereas $\mathbf{X}$ is a matrix). In all cases y is the response variable, ϵ the residual 'error' and other Greek letters indicate model parameters.

a. The 'balanced one-way ANOVA model', $y_{ij} = \beta_i + \epsilon_{ij}$, where $i = 1 \ldots 3$ and $j = 1 \ldots 2$.

b. A model with two explanatory variables: a factor variable and a continuous variable, x:

$$y_i = \beta_j + \gamma x_i + \epsilon_i \quad \text{if obs. } i \text{ is from factor level } j$$

Assume that $i = 1 \ldots 6$, that the first two observations are for factor level 1 and the remaining four for factor level 2, and that the x_i's are 0.1, 0.4, 0.5, 0.3, 0.4 and 0.7.

c. A model with two explanatory factor variables and only one observation per combination of factor variables: $y_{ij} = \alpha + \beta_i + \gamma_j + \epsilon_{ij}$. The first factor ($\beta$) has three levels and the second factor has four levels.

7.6 A statistician has fitted two alternative models to response data y_i. The first is $y_i = \beta_0 + \beta_1 x_i + \epsilon_i$ and the second is $y_i = \beta_0 + \beta_1 x_i + \gamma_j + \epsilon_i$ if y_i from group j. In R the factor variable containing the group labels is `trt`. The statistician wants to test the null hypothesis that the simpler model is correct. To do this, both models are fitted in R, and a fragment of the summary for each is shown here:

```
> summary(b0)
lm(formula = y ~ x)
...
Resid standard error: 0.3009 on 98 degrees of freedom
> summary(b1)
lm(formula = y ~ x + trt)
...
Resid standard error: 0.3031 on 95 degrees of freedom
```

In R this test could be conducted via `anova(b0,b1)`, but instead perform the test using just the information given (and `pf` in R).

7.7 Consider the `cars` model of Section 7.2.2. This question is about the mechanics of estimating that model using the QR decomposition.

a. Create a model matrix $\mathbf{X}$, for the model, using the data in the `cars` data frame and the `model.matrix`.
b. Now form the QR decomposition of $\mathbf{X}$ as follows

```
qrx <- qr(X) ## returns a QR decomposition object
Q <- qr.Q(qrx,complete=TRUE) ## extract Q
R <- qr.R(qrx) ## extract R
```

c. Look at `R` to confirm its structure. Confirm that `Q` is an orthogonal matrix. Confirm that $\|\mathbf{Qx}\|^2 = \|\mathbf{x}\|^2$ for any $\mathbf{x}$ of appropriate dimension, by trying some example $\mathbf{x}$'s (why does this happen?).
d. Obtain $\mathbf{f}$ and $\mathbf{r}$ (in the notation of Section 7.1.1).
e. Evaluate $\hat{\boldsymbol{\beta}}$ using $\mathbf{R}$ and $\mathbf{f}$.
f. Confirm that $\|\mathbf{r}\|^2 = \|\mathbf{y} - \mathbf{X}\hat{\boldsymbol{\beta}}\|^2$ for this model.
g. Estimate σ^2 as $\hat{\sigma}^2 = \|\mathbf{r}\|^2/(n-p)$.
h. Using $\hat{\sigma}^2$ and $\mathbf{R}$, obtain an estimate of the estimator covariance matrix $\mathbf{V}_{\hat{\boldsymbol{\beta}}}$ corresponding to $\hat{\boldsymbol{\beta}}$.

7.8 A material of unknown volume is divided into four roughly equal parts by cutting it in two first and then cutting the resulting pieces in two. Two alternative methods of estimating the (different) volumes of each part are suggested

A. Make two estimates of the volume of each section.
B. Make two estimates of the volume of each of the 2 parts formed first, and one estimate of the volume of the each of the 4 final pieces.

Assuming that each estimate is independent and unbiased with variance σ^2, show that the variances of the least squares estimates of the four volumes are $0.5\sigma^2$ by method A and $0.6\sigma^2$ by method B. Hint: use the $(\mathbf{X}^T\mathbf{X})^{-1}\sigma^2$ form of the parameter estimator covariance matrix.

7.9 A distillery sets up and sponsors a hill race dubbed the 'Whisky Challenge', for promotional purposes. To generate extra interest from elite fell runners in the first year, it is proposed to offer a prize for every runner who completes the course in less than a set time, T_0. The organisers need to set T_0 high enough to generate a big field of participants, but low enough that they do not bankrupt the distillery. To this end they approach you to come up with a predicted winning time for the race. To help you do this, the `hills` data frame in R package `MASS` provides winning times for 35 Scottish hill races. To load the data and examine it, type `library(MASS);hills` in R. Find and estimate a suitable linear model for predicting winning `times` (minutes) in terms of race distance `dist` (miles) and the total height climbed `climb` (feet). If you are not sure where to start look up 'Naismith's Rule' on Wikipedia. The Whisky Challenge is to be a 7-mile race, with 2400 feet of ascent.

Appendix A

Some distributions

This appendix covers some standard distributions useful in the construction of models. The *gamma function* occurs frequently and is defined as

$$\Gamma(x) = \int_0^\infty t^{x-1} e^{-t} dt.$$

See `?gamma` in R to evaluate it numerically. Note that if n is a positive integer then $\Gamma(n) = (n-1)!$

It is also convenient to define the *beta function*,

$$\mathcal{B}(a,b) = \int_0^1 t^{a-1}(1-t)^{b-1} dt = \frac{\Gamma(a)\Gamma(b)}{\Gamma(a+b)},$$

where $a > 0$ and $b > 0$ (both real). See `?beta` in R.

A.1 Continuous random variables: the normal and its relatives

The normal distribution is ubiquitous in statistics, so we start with it and its relatives.

A.1.1 Normal distribution

A random variable X follows a normal (or 'Gaussian') distribution with mean μ and variance σ^2 if it has probability density function

$$f(x) = \frac{1}{\sqrt{2\pi}\sigma} \exp\left\{-\frac{1}{2\sigma^2}(x-\mu)^2\right\}, \quad -\infty < x < \infty,$$

$\sigma^2 > 0$, but μ is unrestricted. Standard notation is $X \sim N(\mu, \sigma^2)$. The central limit theorem of Section 1.9 ensures the central place of the normal distribution in statistics, both as a limiting distribution of estimators and as a reasonable model for many variables that can be characterised as the sum of other random variables. The multivariate normal is equally important, and is characterised in Section 1.6 (see Section 6.5.4 or B.2 for its generation). $Z \sim N(0,1)$ is a *standard normal* random variable. See `?dnorm` in R. Continuous positive variables are often modelled as being log-normally distributed; that is, the logarithm of the variable is assumed normally distributed. See `?dlnorm` in R.

In Bayesian statistics it is often convenient to parameterise the normal in terms of the *precision*, $\tau = 1/\sigma^2$. The normal distribution is conjugate for μ and the gamma distribution is conjugate for τ.

A.1.2 χ^2 distribution

Let $Z_1, Z_2, \ldots, Z_n$ be a set of independent $N(0,1)$ random variables. Then $X = \sum_{i=1}^n Z_i^2$ is a χ_n^2 random variable, with p.d.f.

$$f(x) = \frac{1}{2\Gamma(n/2)} \left(\frac{x}{2}\right)^{n/2-1} e^{-x/2}, \quad x \geq 0.$$

Standard notation is $X \sim \chi_n^2$. $E(X) = n$ and $\text{var}(X) = 2n$. χ^2 random variables often occur when sums of squares of random variables are involved. The χ_2^2 distribution is also the exponential distribution with $\lambda = 1/2$, and if $U \sim U(0,1)$ then $-2\log(U) \sim \chi_2^2$. Notice that the distribution is also defined for non-integer n. See `?dchisq` in R.

A.1.3 t and Cauchy distributions

Let $Z \sim N(0,1)$ and independently $X \sim \chi_n^2$. Then $T = Z/\sqrt{X/n}$ has a t distribution with n degrees of freedom. In short $T \sim t_n$. The p.d.f. is

$$f(t) = \frac{\Gamma(n/2+1/2)}{\sqrt{n\pi}\Gamma(n/2)}(1+t^2/n)^{-n/2-1/2}, \quad -\infty < t < \infty,$$

and $n \geq 1$ need not be an integer. $E(T) = 0$ if $n > 1$, but is otherwise undefined. $\text{var}(T) = n/(n-2)$ for $n > 2$, but is infinite otherwise. t_∞ is $N(0,1)$, whereas for $n < \infty$ the t_n distribution is 'heavier tailed' than a standard normal. t_1 is also known as the *Cauchy* distribution. See Section 1.6.1 for a multivariate version and Section 2.7 for applications. In R see `?dt` and `?dcauchy`.

A.1.4 F distributions

Let $X_n \sim \chi_n^2$ and independently $X_m \sim \chi_m^2$.

$$F = \frac{X_n/n}{X_m/m}$$

has an F distribution with n and m degrees of freedom. In short $F \sim F_{n,m}$. The p.d.f. is

$$f(x) = \frac{\Gamma(n/2 + m/2)n^{n/2}m^{m/2}}{\Gamma(n/2)\Gamma(m/2)} \frac{x^{n/2-1}}{(m + nx)^{n/2+m/2}}, \quad x \geq 0.$$

$E(F) = m/(m-2)$ if $m > 2$. The square root of an $F_{1,n}$ r.v. has a t_n distribution. The F distribution is central to hypothesis testing in linear models and as an approximate reference distribution when using generalised likelihood ratio tests in the presence of nuisance scale parameters (see Section 2.7). See `?df` in R.

A.2 Other continuous random variables

Moving beyond relatives of the normal distribution, two classes of random variable are particularly important: non-negative random variables, and those defined on the unit interval.

A.2.1 Beta distribution

A random variable, X, defined on the unit interval, $[0, 1]$, has a beta distribution if its p.d.f. has the form

$$f(x) = \frac{x^{\alpha-1}(1-x)^{\beta-1}}{\mathcal{B}(\alpha, \beta)},$$

where $\alpha > 0$ and $\beta > 0$ are shape parameters. Standard notation is $X \sim \text{Beta}(\alpha, \beta)$ (although others are used as well):

$$E(X) = \frac{\alpha}{\alpha + \beta} \quad \text{and} \quad \text{var}(X) = \frac{\alpha\beta}{(\alpha+\beta)^2(\alpha+\beta+1)}.$$

In R see `?dbeta`: `shape1` is α and `shape2` is β. The beta distribution is often used as a prior for probabilities (where it is often conjugate). $\text{Beta}(1, 1)$ is the uniform distribution on $[0, 1]$ (so $f(x) = 1$). For the uniform, which can be defined over any finite interval, see `?dunif` in R.

A.2.2 Gamma and exponential distributions

A positive random variable X has a gamma distribution with shape parameter $\alpha > 0$ and scale parameter $\theta > 0$ if its p.d.f. is

$$f(x) = \frac{x^{\alpha-1}e^{-x/\theta}}{\theta^\alpha\Gamma(\alpha)}.$$

$E(X) = \alpha\theta$ and var$(X) = (\alpha\theta^2)$. Standard notation is $X \sim$ Gamma (α, θ), but beware that the distribution is often written in terms of the rate parameter $\beta = 1/\theta$ or even directly in terms of the mean of X and a scale parameter. See `?dgamma`, where `shape` is α and `scale` is θ. In JAGS `gamma` is parameterised using α and β, in that order.

The gamma$(1, \lambda^{-1})$ is the *exponential distribution*, for which $X \geq 0$ (i.e. zero is possible). Its p.d.f. simplifies to $f(x) = \lambda \exp(-\lambda x)$, while $E(X) = \lambda^{-1}$ and var$(X) = \lambda^{-2}$. It is useful for describing the time between independent random events. See `?dexp` in R.

A.2.3 Weibull distribution

Random variable, T, has a Weibull distribution if its p.d.f. is

$$f(t) = \frac{k}{\lambda}\left(\frac{t}{\lambda}\right)^{k-1} e^{-t^k/\lambda^k}, \quad t \geq 0$$

and 0 otherwise. $k > 0$ and $\lambda > 0$. This distribution is often used to model time-to-failure (or other event) data, in which case the failure (event) rate (also known as the hazard function) is given by $k/\lambda(t/\lambda)^{k-1}$.

$$E(T) = \lambda\Gamma(1+1/k) \text{ and } \text{var}(T) = \lambda^2\left\{\Gamma(1+2/k) - \Gamma(1+1/k)^2\right\}.$$

In R see `?dweibull` where k is `shape` and λ is `scale`.

A.2.4 Dirichlet distribution

Consider $\mathbf{X}$, an n vector of non-negative random variables, X_i, where $\sum_{i=1}^n X_i = 1$. $\mathbf{X}$ has a Dirichlet distribution with parameters $\alpha_1, \ldots \alpha_n$ if its p.d.f. is

$$f(\mathbf{x}) = \frac{\Gamma(\alpha_0)}{\prod_{i=1}^n \Gamma(\alpha_i)} \prod_{i=1}^n x_i^{\alpha_i - 1},$$

where $\alpha_0 = \sum_{i=1}^n \alpha_i$.

$$E(X_i) = \frac{\alpha_i}{\alpha_0} \text{ and } \text{var}(X_i) = \frac{\alpha_i(\alpha_0 - \alpha_i)}{\alpha_0^2(\alpha_0 + 1)}.$$

The Dirichlet distribution is typically used as a prior distribution for vectors of probabilities that must sum to 1.

A.3 Discrete random variables

A.3.1 Binomial distribution

Consider n independent trials each with probability of success p. The total number of successes, $x = 0, 1, \ldots, n$, follows a binomial distribution with probability function,

$$f(x) = \frac{n!}{x!(n-x)!} p^x (1-p)^{n-x}.$$

$E(X) = np$ and $\text{var}(X) = np(1-p)$. See `?dbinom` in R.

A.3.2 Poisson distribution

Letting the binomial $n \to \infty$ and $p \to 0$, while holding their product constant at $np = \lambda$, yields the Poisson distribution, with probability mass function

$$f(x) = \frac{\lambda^x \mathrm{e}^{-\lambda}}{x!},$$

where x can be any non-negative integer. $E(X) = \text{var}(X) = \lambda$. The Poisson distribution is often used for count data that can be thought of as counts of rare events. See `?dpois` in R.

A.3.3 Negative binomial distribution

Often count data show higher variance than is consistent with a Poisson distribution, but if we allow the Poisson parameter, λ, to itself have a gamma distribution, then we end up with a more dispersed count distribution: the negative binomial. $X \sim \text{NB}(n, p)$ if its probability function is

$$f(x) = \frac{\Gamma(x+n)}{\Gamma(n)x!}(1-p)^x p^n,$$

where x is a non-negative integer. $E(X) = n(1-p)/p$ and $\text{var}(X) = n(1-p)p^2$. If $n \to \infty$ and $p \to 1$ in such a way that $E(X)$ is held constant at λ, then this tends to $\text{Poi}(\lambda)$. See `?dnbinom` in R. In JAGS the parameter n is known as r.

A.3.4 Hypergeometric distribution

This distribution is useful in problems involving sampling without replacement. Suppose you have an urn containing m white balls and n black balls, and you sample k balls from the urn randomly and without replacement. The number of white balls drawn follows a hypergeometric distribution. See `?dhyper` in R.

A.3.5 Geometric distribution

Consider a sequence of independent trials, each with probability p of success. If X is the number of failures before the first success, then it follows a *geometric* distribution, with probability function

$$f(x) = p(1-p)^x$$

for non-negative integers x. $E(X) = (1-p)/p$ while $\text{var}(X) = (1-p)/p^2$. See `?dgeom` in R.

Appendix B

Matrix computation

Statistical computation often involves numerical computation with matrices. It is quite easy to get this wrong, producing code that is orders of magnitude less efficient or less stable than could be achieved. This appendix introduces the basics of stability and efficiency in matrix computation, along with some standard matrix decompositions useful in statistics. See Golub and Van Loan (2013) or Watkins (1991) for more detail.

B.1 Efficiency in matrix computation

Consider this simple example in R.

```
n <- 2000
A <- matrix(runif(n*n),n,n)
B <- matrix(runif(n*n),n,n)
y <- runif(n)
system.time(f0 <- A%*%B%*%y) ## case 1
  user system elapsed
 31.50  0.03 31.58
system.time(f1 <- A%*%(B%*%y)) ## case 2
  user system elapsed
  0.08  0.00  0.08
```

`f0` and `f1` are identical to machine precision, but `f1` took much less time to compute. Why? The answer is to do with how the multiplications were ordered in the two cases, and the number of *floating point operations* (flops) required by the two alternative orderings.

1. In the first case $\mathbf{AB}$ was formed first, and the resulting matrix used to pre-multiply the vector $\mathbf{y}$.
2. In the second case, the vector $\mathbf{By}$ was formed first and was then pre-multiplied by $\mathbf{A}$.

The first case took more time because the number of floating point operations (+, -, *, /) required to multiply `A` and `B` was about $2n^3$, whereas the number of flops required to multiply an $n \times n$ matrix by an n vector is only about $2n^2$. Hence `f0` cost about $2n^3 + 2n^2$ operations, whereas `f1` cost only $4n^2$. So the first alternative required about $n/2$ times as many operations as the second.[1]

Another simple matrix operation is the evaluation of the trace of a matrix product. Again different ways of computing the same quantity lead to radically different computation times. Consider the computation of $\text{tr}(\mathbf{AB})$ where $\mathbf{A}$ is 5000×100 and $\mathbf{B}$ is 100×5000:

```
n <- 5000;m <- 100
A <- matrix(runif(n*m),n,m)
B <- matrix(runif(n*m),m,n)
system.time(sum(diag(A%*%B)))
 user system elapsed
10.46   0.11  10.58
system.time(sum(diag(B%*%A)))
 user system elapsed
  0.2    0.0    0.2
system.time(sum(A*t(B)))
 user system elapsed
 0.02   0.00   0.02
```

1. The first method forms $\mathbf{AB}$, at a flop cost of $2n^2m$ and then extracts the leading diagonal and sums it.
2. The second method uses the fact that $\text{tr}(\mathbf{AB}) = \text{tr}(\mathbf{BA})$. It forms $\mathbf{BA}$ at flop cost of $2nm^2$ and then extracts the leading diagonal of the result and sums it.
3. The third method makes direct use of the fact that $\text{tr}(\mathbf{AB}) = \sum_{ij} A_{ij}B_{ji}$, at a cost of $2nm$. It is the fastest because no effort is wasted in calculating unused off-diagonal matrix elements.

Notice that method 1 is not just wasteful of flops, but also requires storage of an $n \times n$ matrix, which is much larger than either $\mathbf{A}$ or $\mathbf{B}$.

Unfortunately, it is not generally possible to automatically choose the most efficient alternative when computing with matrices. Even the apparently simple problem of choosing the best order in which to perform matrix multiplication is difficult to automate. However, for many statistical computing problems, a small amount of effort in identifying the best ordering at the coding stage pays big dividends in terms of improved computing speeds. In general, the important thing is to be constantly aware of the

[1] The ratio of timings observed is not exactly 1000 here, because R also spends time interpreting the instructions, setting up storage and so on, which is a significant part of the cost of producing `f1`; low level optimisation also has an impact on timing.

potential for careless coding with matrices to be hugely inefficient, and to consider flop count in all matrix computations.

Generally when flop counting it is important to know whether the count is proportional to n^2 or to n^3 (for example), as $n \to \infty$, but much less important to know whether it is $2n^2$ or $4n^2$, say. For this reason, it is often sufficient to simply consider how the cost scales with the problem size, in the $n \to \infty$ limit, without worrying about exact constants of proportionality. So we simply consider whether an algorithm is $O(n^2)$ or $O(n^3)$ ('order n^2 or order n^3'), for example.

B.2 Choleski decomposition: a matrix square root

Positive definite matrices are the 'positive real numbers' of matrix algebra. They have particular computational advantages and occur frequently in statistics, because covariance matrices are usually positive definite (and always positive semi-definite). So let's start with positive definite matrices, and their matrix square roots. To see why matrix square roots might be useful, consider the following.

Example Generating multivariate normal random variables. There exist very quick and reliable methods for simulating i.i.d. $N(0, 1)$ random deviates, but suppose that $N(\boldsymbol{\mu}, \boldsymbol{\Sigma})$ random vectors are required. Clearly we can generate vectors $\mathbf{z}$ from $N(\mathbf{0}, \mathbf{I})$. If we could find a matrix $\mathbf{R}$ such that $\mathbf{R}^{\mathrm{T}}\mathbf{R} = \boldsymbol{\Sigma}$, then $\mathbf{y} \equiv \mathbf{R}^{\mathrm{T}}\mathbf{z} + \boldsymbol{\mu} \sim N(\boldsymbol{\mu}, \boldsymbol{\Sigma})$, because the covariance matrix of $\mathbf{y}$ is $\mathbf{R}^{\mathrm{T}}\mathbf{I}\mathbf{R} = \mathbf{R}^{\mathrm{T}}\mathbf{R} = \boldsymbol{\Sigma}$ and $E(\mathbf{y}) = E(\mathbf{R}^{\mathrm{T}}\mathbf{z} + \boldsymbol{\mu}) = \boldsymbol{\mu}$.

In general the square root of a positive definite matrix is not uniquely defined, but there is a unique *upper triangular* square root of any positive definite matrix: its *Choleski factor*. The algorithm for finding the Choleski factor is easily derived. Consider a 4×4 example first. The defining matrix equation is

$$\begin{bmatrix} R_{11} & 0 & 0 & 0 \\ R_{12} & R_{22} & 0 & 0 \\ R_{13} & R_{23} & R_{33} & 0 \\ R_{14} & R_{24} & R_{34} & R_{44} \end{bmatrix} \begin{bmatrix} R_{11} & R_{12} & R_{13} & R_{14} \\ 0 & R_{22} & R_{23} & R_{24} \\ 0 & 0 & R_{33} & R_{34} \\ 0 & 0 & 0 & R_{44} \end{bmatrix}$$
$$= \begin{bmatrix} A_{11} & A_{12} & A_{13} & A_{14} \\ A_{12} & A_{22} & A_{23} & A_{24} \\ A_{13} & A_{23} & A_{33} & A_{34} \\ A_{14} & A_{24} & A_{34} & A_{44} \end{bmatrix}.$$

If the component equations of this expression are written out and solved in the right order, then each contains only one unknown, as the following illustrates (unknowns are in bold):

$$\begin{aligned}
A_{11} &= \mathbf{R_{11}}^2 \\
A_{12} &= R_{11}\mathbf{R_{12}} \\
A_{13} &= R_{11}\mathbf{R_{13}} \\
A_{14} &= R_{11}\mathbf{R_{14}} \\
A_{22} &= R_{12}^2 + \mathbf{R_{22}}^2 \\
A_{23} &= R_{12}R_{13} + R_{22}\mathbf{R_{23}} \\
&\;\;\vdots
\end{aligned}$$

Generalising to the $n \times n$ case, and using the convention that $\sum_{k=1}^{0} x_i \equiv 0$, we have

$$R_{ii} = \sqrt{A_{ii} - \sum_{k=1}^{i-1} R_{ki}^2}, \text{ and } R_{ij} = \frac{A_{ij} - \sum_{k=1}^{i-1} R_{ki}R_{kj}}{R_{ii}}, \quad j > i.$$

Working through these equations in row order, from row one, and starting each row from its leading diagonal component, ensures that all right-hand-side quantities are known at each step. Choleski decomposition requires $n^3/3$ flops and n square roots. In R it is performed by function `chol`, which calls routines in LAPACK or LINPACK.[2]

Example (continued) The following simulates 1000 random draws from

$$N\left(\begin{bmatrix} 1 \\ -1 \\ 3 \end{bmatrix}, \begin{bmatrix} 2 & -1 & 1 \\ -1 & 2 & -1 \\ 1 & -1 & 2 \end{bmatrix}\right)$$

and checks their observed mean and covariance:

```
V <- matrix(c(2,-1,1,-1,2,-1,1,-1,2),3,3)
mu <- c(1,-1,3)
R <- chol(V)          ## Choleski factor of V
Z <- matrix(rnorm(3000),3,1000) ## 1000 N(0,I) 3-vectors
Y <- t(R)%*%Z + mu        ## 1000 N(mu,V) vectors

## and check that they behave as expected...
rowMeans(Y)               ## observed mu
```

[2] Actually, numerical analysts do not consider the Choleski factor to be a square root in the strict sense, because of the transpose in $\mathbf{A} = \mathbf{R}^\mathrm{T}\mathbf{R}$.

```
[1] 1.0924086 -0.9694124 2.9926779

 (Y-mu)%*%t(Y-mu)/1000    ## observed V
          [,1]      [,2]      [,3]
[1,] 2.066872 -1.039062 1.003980
[2,] -1.039062 2.054408 -0.980139
[3,] 1.003980 -0.980139 1.833971
```

As a second application of the Choleski decomposition, consider evaluating the log likelihood of $\boldsymbol{\mu}$ and $\boldsymbol{\Sigma}$:

$$l = -\frac{n}{2}\log(2\pi) - \frac{1}{2}\log(|\boldsymbol{\Sigma}|) - \frac{1}{2}(\mathbf{y} - \boldsymbol{\mu})^{\mathrm{T}}\boldsymbol{\Sigma}^{-1}(\mathbf{y} - \boldsymbol{\mu}).$$

If we were simply to invert $\boldsymbol{\Sigma}$ to evaluate the final term, it would cost $2n^3$ flops, and we would still need to evaluate the determinant. A Choleski-based approach is much better. It is easy to see that $\boldsymbol{\Sigma}^{-1} = \mathbf{R}^{-1}\mathbf{R}^{-\mathrm{T}}$, where $\mathbf{R}$ is the Choleski factor of $\boldsymbol{\Sigma}$. So the final term in the log likelihood can be written as $\mathbf{z}^{\mathrm{T}}\mathbf{z}$ where $\mathbf{z} = \mathbf{R}^{-\mathrm{T}}(\mathbf{y} - \boldsymbol{\mu})$. Notice that we do not actually need to evaluate $\mathbf{R}^{-\mathrm{T}}$, but simply to solve $\mathbf{R}^{\mathrm{T}}\mathbf{z} = \mathbf{y} - \boldsymbol{\mu}$ for $\mathbf{z}$. To see how this is done, consider a 4×4 case again:

$$\begin{bmatrix} R_{11} & 0 & 0 & 0 \\ R_{12} & R_{22} & 0 & 0 \\ R_{13} & R_{23} & R_{33} & 0 \\ R_{14} & R_{24} & R_{34} & R_{44} \end{bmatrix} \begin{bmatrix} z_1 \\ z_2 \\ z_3 \\ z_4 \end{bmatrix} = \begin{bmatrix} y_1 - \mu_1 \\ y_2 - \mu_2 \\ y_3 - \mu_3 \\ y_4 - \mu_4 \end{bmatrix}.$$

If this system of equations is solved from the top down, then there is only one unknown (shown in bold) at each stage:

$$\begin{aligned} R_{11}\mathbf{z_1} &= y_1 - \mu_1 \\ R_{12}z_1 + R_{22}\mathbf{z_2} &= y_2 - \mu_2 \\ R_{13}z_1 + R_{23}z_2 + R_{33}\mathbf{z_3} &= y_3 - \mu_3 \\ R_{14}z_1 + R_{24}z_2 + R_{34}z_3 + R_{44}\mathbf{z_4} &= y_4 - \mu_4 \end{aligned}$$

The generalisation of this *forward substitution* process to n dimensions is obvious, as is the fact that it costs $O(n^2)$ flops: much cheaper than explicit formation of $\mathbf{R}^{-\mathrm{T}}$, which would involve applying forward substitution to find each column of the unknown $\mathbf{R}^{-\mathrm{T}}$ in the equation $\mathbf{R}^{\mathrm{T}}\mathbf{R}^{-\mathrm{T}} = \mathbf{I}$, at $O(n^3)$ cost.

In R there is a routine `forwardsolve` for doing forward substitution with a lower triangular matrix (and a routine `backsolve` for performing the equivalent *back substitution* with upper triangular matrices). Before using it, we still need to consider $|\boldsymbol{\Sigma}|$. Again the Choleski factor helps. From the general properties of determinants we know that $|\mathbf{R}^\mathrm{T}||\mathbf{R}| = |\boldsymbol{\Sigma}|$, but because $\mathbf{R}$ is triangular $|\mathbf{R}^\mathrm{T}| = |\mathbf{R}| = \prod_{i=1}^n R_{ii}$. So given the Choleski factor, the calculation of the determinant is $O(n)$.

Example The following evaluates the log likelihood of the covariance matrix `V` and mean vector `mu`, from the previous example, given an observed $\mathbf{y}^\mathrm{T} = (1, 2, 3)$:

```
  y <- 1:3; n <- length(y)
  z <- forwardsolve(t(R),y-mu)
  logLik <- -n*log(2*pi)/2-sum(log(diag(R)))-sum(z*z)/2
  logLik
[1] -6.824963
```

Note that Choleski decomposition of a matrix that is not positive definite will fail. Positive semi-definite is no better, because in that case a leading diagonal element of the Choleski factor will become zero, so that computation of the off-diagonal elements on the same row is impossible. Positive semi-definite matrices are reasonably common, so this is a practical problem. For the positive semi-definite case, it is possible to modify the Choleski decomposition by *pivoting*; that is, by reordering the rows and columns of the original matrix so that the zeroes end up at the end of the leading diagonal of the Choleski factor, in rows that are all zero. This is not pursued further here. Rather let us consider a more general matrix decomposition, that provides matrix square roots along with much else.

B.3 Eigen-decomposition (spectral-decomposition)

Any symmetric matrix, $\mathbf{A}$ can be written as

$$\mathbf{A} = \mathbf{U}\boldsymbol{\Lambda}\mathbf{U}^\mathrm{T}, \tag{B.1}$$

where the matrix $\mathbf{U}$ is orthogonal and $\boldsymbol{\Lambda}$ is a diagonal matrix, with i^{th} leading diagonal element λ_i (conventionally $\lambda_i \ge \lambda_{i+1}$). Post-multiplying both sides of the decomposition by $\mathbf{U}$ we have

$$\mathbf{AU} = \mathbf{U}\boldsymbol{\Lambda}.$$

Considering this system one column at a time and writing $\mathbf{u}_i$ for the ith column of $\mathbf{U}$ we have

$$\mathbf{A}\mathbf{u}_i = \lambda_i \mathbf{u}_i.$$

So the λ_i are the eigenvalues of $\mathbf{A}$, and the columns of $\mathbf{U}$ are the corresponding eigenvectors. (B.1) is the *eigen-decomposition* or *spectral decomposition* of $\mathbf{A}$.

State-of-the-art schemes for eigen-decomposition are fairly intricate, but do *not* use the determinant and characteristic equation of $\mathbf{A}$. One practical scheme is as follows: (i) the matrix $\mathbf{A}$ is first reduced to tri-diagonal form using repeated pre- and post-multiplication by simple rank-one orthogonal matrices called *Householder rotations*, and (ii) an iterative scheme called *QR-iteration*[3] then pre-and post-multiplies the tri-diagonal matrix by even simpler orthogonal matrices, in order to reduce it to diagonal form. At this point the diagonal matrix contains the eigenvalues, and the product of all the orthogonal matrices gives $\mathbf{U}$. Eigen-decomposition is $O(n^3)$, but a good symmetric eigen routine is around 10 times as computationally costly as a Choleski routine.

An immediate use of the eigen-decomposition is to provide an alternative characterisation of positive (semi-) definite matrices. All the eigenvalues of a positive (semi-) definite matrix must be positive (non-negative) and real. This is easy to see. Were some eigenvalue, λ_i to be negative (zero), then the corresponding eigenvector $\mathbf{u}_i$ would result in $\mathbf{u}_i^{\mathrm{T}}\mathbf{A}\mathbf{u}_i$ being negative (zero). At the same time the existence of an $\mathbf{x}$ such that $\mathbf{x}^{\mathrm{T}}\mathbf{A}\mathbf{x}$ is negative (zero) leads to a contradiction unless at least one eigenvalue is negative (zero).[4]

B.3.1 Powers of matrices

Consider raising $\mathbf{A}$ to the power m.

$$\begin{aligned}\mathbf{A}^m = \mathbf{A}\mathbf{A}\mathbf{A}\cdots\mathbf{A} = \mathbf{U}\boldsymbol{\Lambda}\mathbf{U}^{\mathrm{T}}\mathbf{U}\boldsymbol{\Lambda}\mathbf{U}^{\mathrm{T}}\cdots\mathbf{U}\boldsymbol{\Lambda}\mathbf{U}^{\mathrm{T}} \\ = \mathbf{U}\boldsymbol{\Lambda}\boldsymbol{\Lambda}\cdots\boldsymbol{\Lambda}\mathbf{U}^{\mathrm{T}} = \mathbf{U}\boldsymbol{\Lambda}^m\mathbf{U}^{\mathrm{T}},\end{aligned}$$

where $\boldsymbol{\Lambda}^m$ is just the diagonal matrix with λ_i^m as the ith leading diagonal element. This suggests that any real valued function, f, of a real valued argument, which has a power series representation, has a natural generalisation to a symmetric matrix valued function of a symmetric matrix argument; that is

$$f'(\mathbf{A}) \equiv \mathbf{U}f'(\boldsymbol{\Lambda})\mathbf{U}^{\mathrm{T}},$$

where $f'(\boldsymbol{\Lambda})$ denotes the diagonal matrix with i^{th} leading diagonal element $f(\lambda_i)$. For example, $\exp(\mathbf{A}) = \mathbf{U}\exp(\boldsymbol{\Lambda})\mathbf{U}^{\mathrm{T}}$.

[3] Not to be confused with QR decomposition.

[4] We can write $\mathbf{x} = \mathbf{U}\mathbf{b}$ for some vector $\mathbf{b}$. So $\mathbf{x}^{\mathrm{T}}\mathbf{A}\mathbf{x} < 0 \Rightarrow \mathbf{b}^{\mathrm{T}}\boldsymbol{\Lambda}\mathbf{b} < 0 \Rightarrow \sum b_i^2\Lambda_i < 0 \Rightarrow \Lambda_i < 0$ for some i.

B.3.2 Another matrix square root

For matrices with non-negative eigenvalues we can generalise to non-integer powers. For example, it is readily verified that $\sqrt{\mathbf{A}} = \mathbf{U}\sqrt{\mathbf{\Lambda}}\mathbf{U}^{\mathrm{T}}$ has the property that $\sqrt{\mathbf{A}}\sqrt{\mathbf{A}} = \mathbf{A}$. Notice (i) that $\sqrt{\mathbf{A}}$ is not the same as the Choleski factor, emphasizing the non-uniqueness of matrix square roots and (ii) that, unlike the Choleski factor, $\sqrt{\mathbf{A}}$ is well defined for positive semi-definite matrices (and can therefore be computed for *any* covariance matrix).

B.3.3 Matrix inversion, rank and condition

Continuing in the same vein we can investigate matrix inversion by writing

$$\mathbf{A}^{-1} = \mathbf{U}\mathbf{\Lambda}^{-1}\mathbf{U}^{\mathrm{T}},$$

where the diagonal matrix $\mathbf{\Lambda}^{-1}$ has i^{th} leading diagonal element λ_i^{-1}. Clearly we have a problem if any of the λ_i are zero, for the matrix inverse will be undefined. A matrix with no zero eigenvalues is termed *full rank*. A matrix with any zero eigenvalues is *rank deficient* and does not have an inverse. The number of nonzero eigenvalues is the *rank* of a matrix.

For some purposes it is sufficient to define a *generalised inverse* or *pseudoinverse* when faced with rank deficiency, by finding the reciprocal of the nonzero eigenvalues, but setting the reciprocal of the zero eigenvalues to zero. This is not pursued here.

It is important to understand the consequences of rank deficiency quite well when performing matrix operations involving matrix inversion/matrix equation solving. This is because near rank deficiency is rather easy to achieve by accident, and in finite precision arithmetic it is as bad as rank deficiency. First consider trying to solve

$$\mathbf{A}\mathbf{x} = \mathbf{y}$$

for $\mathbf{x}$ when $\mathbf{A}$ is rank deficient. In terms of the eigen-decomposition the solution is

$$\mathbf{x} = \mathbf{U}\mathbf{\Lambda}^{-1}\mathbf{U}^{\mathrm{T}}\mathbf{y}.$$

So $\mathbf{y}$ is rotated to become $\mathbf{y}' = \mathbf{U}^{\mathrm{T}}\mathbf{y}$, the elements of $\mathbf{y}'$ are then divided by the eigenvalues, λ_i, and the reverse rotation is applied to the result. The problem is that y_i'/λ_i is not defined if $\lambda_i = 0$. This is just a different way of showing something that you already know: rank-deficient matrices cannot be inverted. But the approach also helps in understanding near rank deficiency and *ill conditioning*.

An illustrative example highlights the problem. Suppose that an $n \times n$ symmetric matrix $\mathbf{A}$ has $n - 1$ distinct eigenvalues ranging from 0.5 to 1, and one much smaller magnitude eigenvalue ϵ. Further suppose that we wish to compute with $\mathbf{A}$ on a machine that can represent real numbers to an accuracy of 1 part in ϵ^{-1}. Now consider solving the system

$$\mathbf{A}\mathbf{x} = \mathbf{u}_1 \tag{B.2}$$

for $\mathbf{x}$, where $\mathbf{u}_1$ is the dominant eigenvector of $\mathbf{A}$. Clearly the correct solution is $\mathbf{x} = \mathbf{u}_1$, but now consider computing the answer. As before we have a formal solution,

$$\mathbf{x} = \mathbf{U}\mathbf{\Lambda}^{-1}\mathbf{U}^{\mathrm{T}}\mathbf{u}_1,$$

but although analytically $\mathbf{u}_1' = \mathbf{U}^{\mathrm{T}}\mathbf{u}_1 = (1, 0, 0, \ldots, 0)^{\mathrm{T}}$, the best we can hope for computationally is to get $\mathbf{u}_1' = (1 + e_1, e_2, e_3, \ldots, e_n)^{\mathrm{T}}$ where the numbers e_j are of the order of $\pm\epsilon$. For convenience, suppose that $e_n = \epsilon$. Then, approximately, $\mathbf{\Lambda}^{-1}\mathbf{u}_1' = (1, 0, 0, \ldots, 0, 1)^{\mathrm{T}}$, and $\mathbf{x} = \mathbf{U}\mathbf{\Lambda}^{-1}\mathbf{u}_1' = \mathbf{u}_1 + \mathbf{u}_n$, which is not correct. Similar distortions would occur if we used any of the other first $n - 1$ eigenvectors in place of $\mathbf{u}_1$: they all become distorted by a spurious $\mathbf{u}_n$ component, with only $\mathbf{u}_n$ itself escaping.

Now consider an *arbitrary* vector $\mathbf{y}$ on the right-hand-side of (B.2). We can always write it as some weighted sum of the eigenvectors $\mathbf{y} = \sum w_i \mathbf{u}_i$. This emphasises how bad the ill-conditioning problem is: all but one of $\mathbf{y}$'s components are seriously distorted when multiplied by $\mathbf{A}^{-1}$. By contrast, multiplication by $\mathbf{A}$ itself would lead only to distortion of the $\mathbf{u}_n$ component of $\mathbf{y}$, and not the other eigenvectors, but the $\mathbf{u}_n$ component is the component that is so heavily shrunken by multiplication by $\mathbf{A}$ that it makes almost no contribution to the result, unless we have the misfortune to choose a $\mathbf{y}$ that is proportional to $\mathbf{u}_n$ and nothing else.

A careful examination of the preceding argument reveals that what really matters in determining the seriousness of the consequences of near rank deficiency is the ratio of the largest magnitude to the smallest magnitude eigenvalues:

$$\kappa = \max |\lambda_i| / \min |\lambda_i|.$$

This quantity is a *condition number* for $\mathbf{A}$.[5] Roughly speaking it is the factor by which errors in $\mathbf{y}$ will be multiplied when solving $\mathbf{A}\mathbf{x} = \mathbf{y}$ for

[5] Because the condition number is so important in numerical computation, there are several methods for getting an approximate condition number more cheaply than via eigen decomposition – e.g. see `?kappa` in R.

$\mathbf{x}$. Once κ begins to approach the reciprocal of the machine precision we are in trouble. A system with a large condition number is referred to as *ill-conditioned*. Orthogonal matrices have $\kappa = 1$, which is why numerical analysts like them so much.

Example Consider a simple simulation in which data are simulated from a quadratic model, and an attempt is made to obtain least squares estimates of the linear model parameters directly from $\hat{\boldsymbol{\beta}} = (\mathbf{X}^{\mathrm{T}}\mathbf{X})^{-1}\mathbf{X}^{\mathrm{T}}\mathbf{y}$.

```
set.seed(1); n <- 100
xx <- sort(runif(n))
y <- .2*(xx-.5)+(xx-.5)^2 + rnorm(n)*.1
x <- xx+100
X <- model.matrix(~ x + I(x^2))
beta.hat <- solve(t(X)%*%X,t(X)%*%y)
Error in solve.default(t(X) %*% X, t(X) %*% y) :
system is computationally singular:
reciprocal condition number = 3.98648e-19
```

This is an apparently innocuous linear model fitting problem. However, the simple fact that the x range is from 100 to 101 has caused the columns of $\mathbf{X}$ to be sufficiently close to linear dependence that $\mathbf{X}^{\mathrm{T}}\mathbf{X}$ is close to singular, as we can confirm by direct computation of its condition number:

```
XtX <- crossprod(X)  ## form t(X)%*%X (efficiently)
lambda <- eigen(XtX)$values
lambda[1]/lambda[3]  ## the condition number of X'X
[1] 2.506267e+18
```

Of course, this raises two obvious questions. Could we have diagnosed the problem directly from $\mathbf{X}$? And how does the `lm` function avoid this problem (it is able to fit this model)? Answers to these questions follow, but first consider a trick for reducing κ.

B.3.4 Preconditioning

The discussion of condition numbers related to systems involving unstructured matrices (albeit *presented* only in the context of symmetric matrices). Systems involving matrices with special structure are sometimes less susceptible to ill-conditioning than naive computation of the condition number would suggest. For example, if $\mathbf{D}$ is a diagonal matrix, then we can accurately solve $\mathbf{Dy} = \mathbf{x}$ for $\mathbf{y}$, however large $\kappa(\mathbf{D})$ is: overflow or underflow are the only limits.

This basic fact can sometimes be exploited to rescale a problem to improve computational stability. As an example consider *diagonal* preconditioning of the computation of $(\mathbf{X}^{\mathrm{T}}\mathbf{X})^{-1}$ considered previously. For $\mathbf{X}^{\mathrm{T}}\mathbf{X}$ we have

```
  solve(XtX)
Error in solve.default(XtX) :
  system is computationally singular:
  reciprocal condition number = 3.98657e-19
```

But now suppose that we create a diagonal matrix $\mathbf{D}$, with elements $D_{ii} = 1/\sqrt{(\mathbf{X}^{\mathrm{T}}\mathbf{X})_{ii}}$. Clearly,

$$(\mathbf{X}^{\mathrm{T}}\mathbf{X})^{-1} = \mathbf{D}(\mathbf{D}\mathbf{X}^{\mathrm{T}}\mathbf{X}\mathbf{D})^{-1}\mathbf{D},$$

but $(\mathbf{D}\mathbf{X}^{\mathrm{T}}\mathbf{X}\mathbf{D})^{-1}$ turns out to have a much lower condition number than $\mathbf{X}^{\mathrm{T}}\mathbf{X}$:

```
  D <- diag(1/diag(XtX)^.5)
  DXXD <- D%*%XtX%*%D
  lambda <- eigen(DXXD)$values
  lambda[1]/lambda[3]
[1] 4.29375e+11
```

As a result we can now compute the inverse of $\mathbf{X}^{\mathrm{T}}\mathbf{X}$:

```
  XtXi <- D%*%solve(DXXD,D) ## computable inverse of X'X
  XtXi %*% XtX              ## how accurate?
       (Intercept)           x      I(x^2)
[1,] 9.999941e-01 -3.058910e-04 0.005661011
[2,] 1.629232e-07 1.000017e+00 0.001764774
[3,] -6.816663e-10 -8.240750e-08 0.999998398
```

This is not perfect, but is better than no answer at all.

B.3.5 Asymmetric eigen-decomposition

If positive definite matrices are the positive reals of the square matrix system, and symmetric matrices are the reals, then asymmetric matrices are the complex numbers. As such they have complex eigenvectors and eigenvalues. It becomes necessary to distinguish right and left eigenvectors (one is no longer the transpose of the other), and the right and left eigenvector matrices are no longer orthogonal matrices (although they are still inverses of each other). Eigen-decomposition of asymmetric matrices is still $O(n^3)$, but is substantially more expensive than the symmetric case. For example, using a basic R setup on a Linux laptop asymmetric eigen-decomposition took four times longer than symmetric eigen-decomposition for a 1000×1000 matrix.

The need to compute with complex numbers somewhat reduces the practical utility of the eigen-decomposition in numerical methods for statistics.

It would be better to have a decomposition that provides some of the useful properties of the eigen-decomposition without the inconvenience of complex numbers. The singular value decomposition (SVD) meets this need.

B.4 Singular value decomposition

The *singular values*, d_i, of an $r \times c$ matrix, $\mathbf{A}$ ($r \geq c$) are the non-negative square roots of the eigenvalues of $\mathbf{A}^{\mathrm{T}}\mathbf{A}$. If $\mathbf{A}$ is positive semi-definite then its singular values are just its eigenvalues, of course. For symmetric matrices, eigenvalues and singular values differ only in sign, if at all. However, the singular values are also well defined and real for matrices that are not even square, let alone symmetric.

Related to the singular values is the *singular value decomposition*,

$$\mathbf{A} = \mathbf{U}\mathbf{D}\mathbf{V}^{\mathrm{T}},$$

where $\mathbf{U}$ has orthogonal columns and is the same dimension as $\mathbf{A}$, while $c \times c$ matrix $\mathbf{D} = \text{diag}(d_i)$ (usually arranged in descending order), and $\mathbf{V}$ is a $c \times c$ orthogonal matrix.

The singular value decomposition is computed using a similar approach to that used for the symmetric eigen problem: orthogonal bi-diagonalization, followed by QR iteration at $O(rc^2)$ cost (it does *not* involve forming $\mathbf{A}^{\mathrm{T}}\mathbf{A}$). It is more costly than symmetric eigen-decomposition, but cheaper than the asymmetric equivalent. For a 1000×1000 matrix, SVD took about 2.5 times as long as symmetric eigen-decomposition using R.

The number of its nonzero singular values gives the rank of a matrix, and the SVD is the most reliable method for numerical rank determination (by examining the size of the singular values relative to the largest singular value). In a similar vein, a general definition of the condition number is the ratio of largest and smallest singular values: $\kappa = d_1/d_c$.

Example Continuing the example of the simple quadratic regression fitting failure, consider the singular values of `X`:

```
 d <- svd(X)$d ## get the singular values of X
 d
[1] 1.010455e+05 2.662169e+00 6.474081e-05
```

Clearly, numerically `X` is close to being rank 2, rather than rank 3. Turning to the condition number,

```
 d[1]/d[3]
[1] 1560769713
```

$\kappa \approx 2\times10^9$ is rather large, especially as it is easy to show that the condition number of $\mathbf{X}^\mathrm{T}\mathbf{X}$ must then be $\kappa^2 \approx 4 \times 10^{18}$. So we now have a pretty clear diagnosis of the cause of the original problem.

In fact the SVD provides not only a diagnosis of the problem, but also one possible solution. We can rewrite the solution to the normal equations in terms of the SVD of X:

$$\begin{aligned}(\mathbf{X}^\mathrm{T}\mathbf{X})^{-1}\mathbf{X}^\mathrm{T}\mathbf{y} &= (\mathbf{VDU}^\mathrm{T}\mathbf{UDV}^\mathrm{T})^{-1}\mathbf{VDU}^\mathrm{T}\mathbf{y}\\ &= (\mathbf{VD}^2\mathbf{V}^\mathrm{T})^{-1}\mathbf{VDU}^\mathrm{T}\mathbf{y}\\ &= \mathbf{VD}^{-2}\mathbf{V}^\mathrm{T}\mathbf{VDU}^\mathrm{T}\mathbf{y}\\ &= \mathbf{VD}^{-1}\mathbf{U}^\mathrm{T}\mathbf{y}\end{aligned}$$

Notice two things:

1. The condition number of the system that we have ended up with is exactly the condition number of $\mathbf{X}$ (i.e. the square root of the condition number involved in the direct solution of the normal equations).
2. Comparing the final right-hand-side expression to the representation of an inverse in terms of its eigen-decomposition, it is clear that $\mathbf{VD}^{-1}\mathbf{U}^\mathrm{T}$ is a sort of *pseudoinverse* of $\mathbf{X}$.

The SVD has many uses. One interesting one is low-rank approximation of matrices. In a well-defined sense, the best rank $k \le \mathrm{rank}(\mathbf{X})$ approximation to a matrix $\mathbf{X}$ can be expressed in terms of the SVD of $\mathbf{X}$ as

$$\tilde{\mathbf{X}} = \mathbf{U}\tilde{\mathbf{D}}\mathbf{V}^\mathrm{T}$$

where $\tilde{\mathbf{D}}$ is $\mathbf{D}$ with all but the k largest singular values set to 0. Using this result to find low-rank approximations to observed covariance matrices is the basis for several dimension-reduction techniques in multivariate statistics (although, of course, a symmetric eigen-decomposition is then equivalent to SVD). One issue with this sort of approximation is that the full SVD is computed, despite the fact that part of it is then to be discarded (be careful with routines that ask you how many eigen or singular vectors to return: I saved 0.1 of a second, out of 13, by getting R routine `svd` to only return the first columns of $\mathbf{U}$ and $\mathbf{V}$). Look up Lanczos methods and Krylov subspaces for approaches that avoid this sort of waste.

B.5 The QR decomposition

The SVD provided a stable solution to the linear model fitting example, but at a rather high computational cost, prompting the question of whether similar stability could be obtained without the full cost of SVD. The QR decomposition provides a positive answer, as was shown in Section 7.1.1. We can write any $r \times c$ rectangular matrix $\mathbf{X}$ ($r \geq c$) as the product of columns of an orthogonal matrix and an upper triangular matrix:

$$\mathbf{X} = \mathbf{QR},$$

where $\mathbf{R}$ is upper triangular and $\mathbf{Q}$ is of the same dimension as $\mathbf{X}$, with orthogonal columns (so $\mathbf{Q}^{\mathrm{T}}\mathbf{Q} = \mathbf{I}$, but $\mathbf{QQ}^{\mathrm{T}} \neq \mathbf{I}$). The QR decomposition has a cost of $O(rc^2)$, but it is about one-third of the cost of SVD. The SVD and QR approaches are the most numerically stable methods for least squares problems, but there is no magic here: it is quite possible to produce model matrices so close to co-linear that these methods also fail. The lesson here is that if possible we should try to set up models so that condition numbers stay low.

Another application of the QR decomposition is determinant calculation. If $\mathbf{A}$ is square and $\mathbf{A} = \mathbf{QR}$ then

$$|\mathbf{A}| = |\mathbf{Q}||\mathbf{R}| = |\mathbf{R}| = \prod_i R_{ii},$$

since $\mathbf{R}$ is triangular, while $\mathbf{Q}$ is orthogonal with determinant 1. Usually we need

$$\log|\mathbf{A}| = \sum_i \log|R_{ii}|,$$

which underflows to $-\infty$ much less easily than $|\mathbf{A}|$ underflows to zero.

B.6 Sparse matrices

Many statistical problems involve *sparse matrices*: matrices that contain a very high proportion of zeroes. This sparsity can be exploited to save on computer memory and floating point operations. We need only store the nonzero entries of a sparse matrix, along with the location of those entries, and need only perform floating point operations when they involve nonzero matrix elements. Many libraries exist for exploiting sparse matrices, such as the `Matrix` package in R. The main difficulty in exploiting sparsity is *infil*: a sparse matrix rarely has a sparse inverse or Choloeski factor, for example, and even the product of two sparse matrices is often not sparse. However, it is the case that a *pivoted* version of a matrix (one in which rows and columns are reordered) has a sparse Choleski factor, for example: so with careful structuring, efficiency can be achieved in some cases. See Davis (2006) for a good introduction.

Appendix C

Random number generation

Chapter 6, in particular, took it for granted that we can produce random numbers from various distributions. Actually we can't. The best that can be done is to produce a completely deterministic sequence of numbers that appears indistinguishable from a random sequence with respect to any relevant statistical property that we choose to test.[1] In other words, we may be able to produce a deterministic sequence of numbers that can be very well modelled as being a random sequence from some distribution. Such deterministic sequences are referred to as sequences of *pseudorandom* numbers, but the *pseudo* part usually gets dropped at some point.

The fundamental problem, for our purposes, is to generate a pseudorandom sequence that can be extremely well modelled as i.i.d. $U(0, 1)$. Given such a sequence, it is fairly straightforward to generate deviates from other distributions, but the i.i.d. $U(0, 1)$ generation is where the problems lie. Indeed if you read around this topic, most books will largely agree about how to turn uniform random deviates into deviates from a huge range of other distributions, but advice on how to obtain the uniform deviates in the first place is much less consistent.

C.1 Simple generators and what can go wrong

Since the 1950s there has been much work on linear congruential generators. The intuitive motivation is something like this. Suppose I take an integer, multiply it by some enormous factor, rewrite it in base – 'something huge', and then throw away everything except for the digits after the decimal point. Pretty hard to predict the result, no? So, if I repeat the operation, feeding each step's output into the input for the next step, a more or

[1] Hence the interesting paradox that although statistical methods in general may be viewed as methods for distinguishing between the deterministic and the random, many statistical methods rely fundamentally on the inability to distinguish random from deterministic.

less random sequence might result. Formally the pseudorandom sequence is defined by

$$X_{i+1} = (aX_i + b)\text{mod}M,$$

where b is 0 or 1, in practice. This is started with a *seed* X_0. The X_i are integers ($< M$, of course), but we can define $U_i = X_i/M$. Now the intuitive hope that this recipe might lead to U_i that are reasonably well modelled by i.i.d. $U(0,1)$ r.v.s is only realized for some quite special choices of a and M, and it takes some number theory to give the generator any sort of theoretical foundation (see Ripley, 1987, Chapter 2).

An obvious property to try to achieve is *full period*. We would like the generator to visit all possible integers between $1-b$ and $M-1$ once before it starts to repeat itself (clearly the first time it revisits a value, it starts to repeat itself). We would also like successive U_is to appear uncorrelated. A notorious and widely used generator called RANDU, supplied at one time with IBM machines, met these basic considerations with

$$X_{i+1} = (65539X_i)\text{mod}2^{31}.$$

This appears to do very well in 1 dimension.

```
n <- 100000 ## code NOT for serious use
x <- rep(1,n)
a <- 65539;M <- 2^31;b <- 0 ## Randu
for (i in 2:n) x[i] <- (a*x[i-1]+b)%%M
u <- x/(M-1)
qqplot((1:n-.5)/n,sort(u))
```

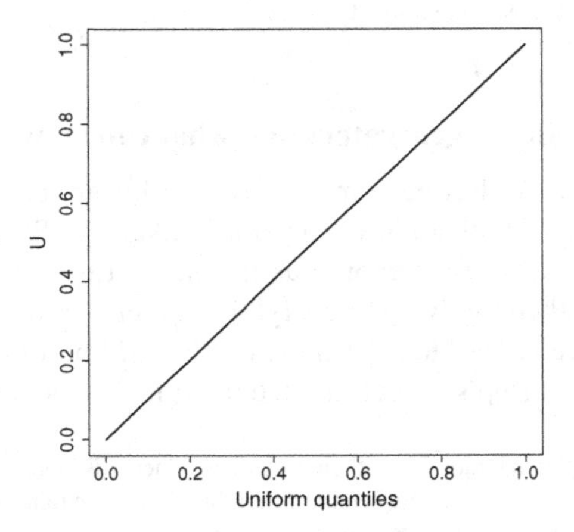

Similarly a plot of U_i vs U_{i-1} indicates no worries with serial correlation:

```
## Create data frame with U at 3 lags...
U <- data.frame(u1=u[1:(n-2)],u2=u[2:(n-1)],u3=u[3:n])
plot(U$u1,U$u2,pch=".")
```

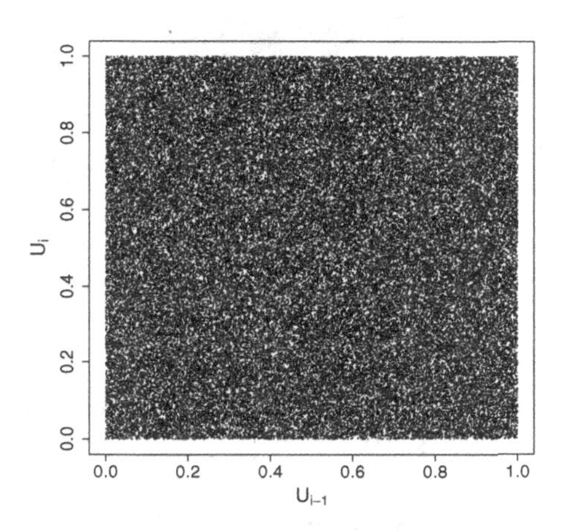

We can also check visually what the distribution of the triples (U_i, U_{i-1}, U_{i-2}) looks like:

```
library(lattice)
cloud(u1~u2*u3,U,pch=".",col=1,screen=list(z=40,x=-70,y=0))
```

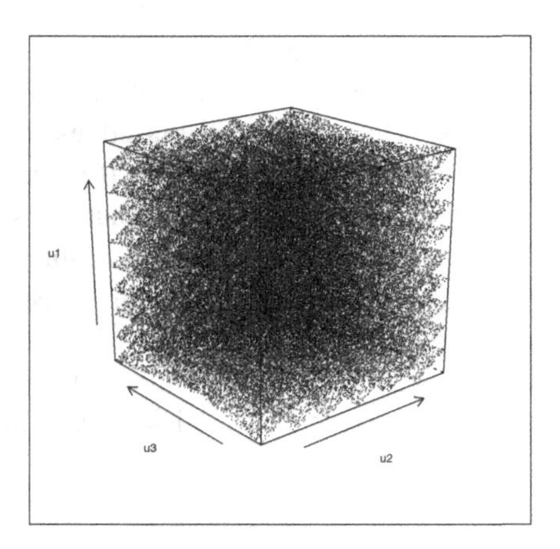

Clearly not quite so random looking. Experimenting a little with rotations gives:

```
cloud(u1~u2*u3,U,pch=".",col=1,screen=list(z=40,x=70,y=0))
```

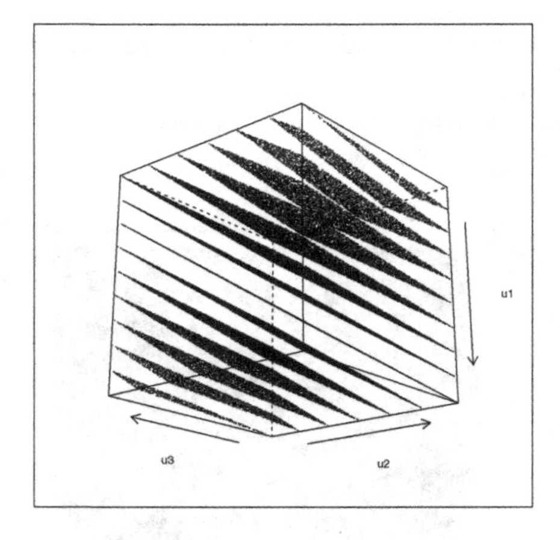

The triples lie on one of 15 planes. Actually it can be shown that this must happen (see Ripley, 1987, §2.2).

Does this deficiency matter in practice? Much of the statistical use of random numbers is for processes somehow equivalent to high dimensional integration. Statistical estimators of integrals have nonzero variance, but can be designed to be unbiased, and this unbiasedness is usually not affected by the dimension of the integral. Deterministic attempts at integration tend to evaluate the integrand on a discrete lattice. They have no variance, but their bias is determined by the lattice spacing, and for fixed computational effort this bias increases sharply with dimension. As a result, statistical estimation of high-dimensional integrals usually outperforms deterministic quadrature rules. However, the unbiasedness of such estimators relies on being able to generate random numbers. If we are actually generating numbers on a lattice, then there is a danger that our statistical estimators may suffer from the same bias problems as deterministic integration.

So the first lesson is to use generators that have been carefully engineered by people with a good understanding of number theory and have then been empirically tested (Marsaglia's *Diehard* battery of tests provides one standard test set). For example, if we stick with simple congruential generators, then

$$X_i = (69069X_{i-1} + 1)\text{mod}2^{32} \tag{C.1}$$

is a much better bet. Here is its triples plot, for which no amount of rotation provides any evidence of structure:

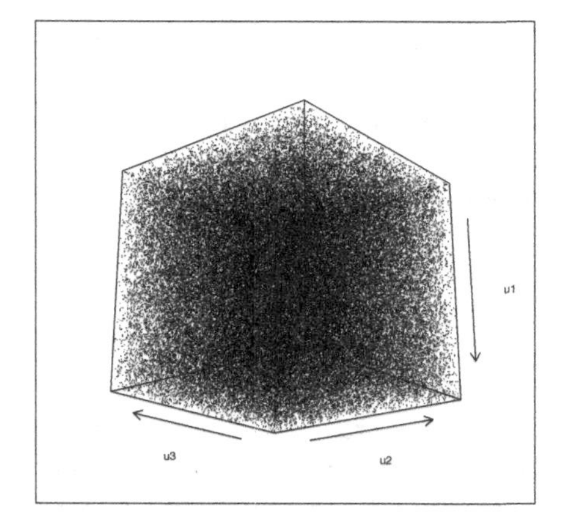

Although this generator is *much* better than RANDU, it is still problematic. An obvious infelicity is the fact that a *very* small X_i will always be followed by an unusually small X_{i+1} (consider $X_i = 1$, for example). This is not a property that would be desirable in a time series simulation, for example. Not quite so obvious is the fact that for any congruential generator of period M, then k-tuples, $U_i, U_{i-1}, \ldots, U_{i-k+1}$ will tend to lie on a finite number of $k-1$ dimensional planes (e.g. for RANDU we saw 3-tuples lying on two dimensional planes.) There will be *at most* $M^{1/k}$ such planes, and as RANDU shows, there can be far fewer. The upshot of this is that if we could visualize 8 dimensions, then the 8-tuple plot for (C.1) would be just as alarming as the 3D plot was for RANDU. Eight is not an unreasonably large dimension for an integral.

Generally then, it would be nice to have generators with better behaviour than simple congruential generators, and in particular we would like generators where k-tuples appear uniformly distributed on $[0, 1]^k$ for as high a k as possible (referred to as having a high *k-distribution*).

C.2 Building better generators

An alternative to the congruential generators are generators that focus on generating random sequences of 0s and 1s. In some ways this seems to be the natural *fundamental* random number generation problem when using modern digital computers, and at the time of writing it also seems to be the approach that yields the most satisfactory results. Such generators are often termed *shift-register* generators. The basic approach is to use bitwise binary operations to make a binary sequence 'scramble itself'. An example is the Marsaglia (2003) *Xorshift* generator as recommended in Press et al. (2007).

Let x be a 64-bit variable (i.e. an array of 64 0s or 1s). The generator is initialised by setting to any value (other than 64 0s). The following steps then constitute one iteration (update of x):

$$x \leftarrow x \wedge (x >> a)$$
$$x \leftarrow x \wedge (x << b)$$
$$x \leftarrow x \wedge (x >> c)$$

Each iteration generates a new random sequence of 0s and 1s. $\wedge$ denotes 'exclusive or' (XOR), and $>>$ and $<<$ are right-shift and left shift, respectively, with the integers a, b and c giving the distance to shift. $a = 21$, $b = 35$ and $c = 4$ appear to be good constants (but see Press et al., 2007, for some others).

If you are a bit rusty on these binary operators then consider an 8-bit example where `x=10011011` and `z=01011100`:

- `x<<1` is `00110110`: the bit pattern is shifted leftwards, with the leftmost bit discarded, and the rightmost set to zero.
- `x<<2` is `01101100`: the pattern is shifted 2 bits leftwards, which also entails discarding the 2 leftmost bits and zeroing the two rightmost.
- `x>>1` is `01001101`: shift the pattern 1 bit rightwards.
- `x^z` is `11000111`:a 1 where the bits in `x` and `z` disagree, and a 0 where they agree.

The Xorshift generator is very fast, has a period of $2^{64}-1$, and passes the Diehard battery of tests (perhaps unsurprising as Marsaglia is responsible for that too). These shift-register generators suffer similar granularity problems to congruential generators (there is always some k for which $[0, 1]^k$ cannot be very well covered by even $2^{64} - 1$ points), but tend to have all bit positions 'equally random', whereas lower order bits from congruential generator sequences often have a good deal of structure.

Now we reach a fork in the road. To achieve better performance in terms of longer period, larger k-distribution, and fewer low-order correlation problems, there seem to be two main approaches: the first pragmatic, and the second more theoretical.

1. Combine the output from several 'good', well-understood, simple generators using operations that maintain randomness (e.g. XOR and addition, but not multiplication). When doing this, the output from the combined generators is *never* fed back into the driving generators.

Preferably combine rather different types of generator. Press et al. (2007) make a convincing case for this approach. Wichmann and Hill (1982), available in R, is an example of such a combined generator, albeit based on three very closely related generators.

2. Use more complicated generators: non-linear or with a higher dimensional state that just a single X_i (see Gentle, 2003). For example, use a shift-register type generator based on maintaining the history of the last n bit-patterns, and using these in the bit scrambling operation. The Matsumoto and Nishimura (1998) *Mersenne Twister* is of this type. It achieves a period of $2^{19937} - 1$ (that is not a misprint: $2^{19937} - 1$ is a 'Mersenne prime'[2]), and is 623-distributed at 32 bit accuracy. That is, its 623-tuples appear uniformally distributed (each appearing the same number of times in a full period) and are spaced 2^{-32} apart (without the ludicrous period this would not be possible). It passes the Diehard tests, is the default generator in R, and C source code is freely available.

C.3 Uniform generation conclusions

This brief discussion shows that random number generation and use of pseudorandom numbers, are nontrivial topics that require some care. That said, most of the time, provided you pick a good modern generator, you will probably have no problems. As general guidelines:

1. Avoid using black-box routines supplied with low level languages such as C: you do not know what you are getting, and there is a history of these being botched.
2. Do make sure you know what method is being used to generate any uniform random deviates that you use and that you are satisfied that it is good enough for your purposes.
3. For any random number generation task that relies on k-tuples having uniform distribution for high k, be particularly careful about what generator you use. This includes any statistical task that is somehow equivalent to high-dimensional integration.
4. The Mersenne Twister is probably the sensible default choice in most cases at present. For high-dimensional problems it remains a good idea to check answers with a different high-quality generator. If results differ significantly, then you will need to find out why (probably starting with the 'other' generator).

[2] Numbers this large are often described as being 'astronomical', but this does not really do it justice: there are probably fewer than 2^{270} atoms in the universe.

Note that I have not discussed methods used by cryptographers. Cryptographers want to use (pseudo)random sequences of bits (0s and 1s) to scramble messages. Their main concern is that if someone were to intercept the random sequence and guess the generator being used, that individual should not be able to infer the state of the generator. Achieving this goal is quite computer intensive, which is why generators used for cryptography are usually over-engineered for simulation purposes.

C.4 Other deviates

Once you have a satisfactory stream of i.i.d. $U(0, 1)$ deviates, then generating deviates from other standard distributions is much more straightforward. Conceptually, the simplest approach is inversion. We know that if X is from a distribution with continuous c.d.f. F, then $F(X) \sim U(0, 1)$. Similarly, if we define the inverse of F by $F^{-}(u) = \min(x|F(x) \geq u)$, and if $U \sim U(0, 1)$, then $F^{-}(U)$ has a distribution with c.d.f. F (this time with not even a continuity restriction on F itself).

As an example here is inversion used to generate one million i.i.d. $N(0, 1)$ deviates in R:

```
system.time(X <- qnorm(runif(1e6)))
 user system elapsed
 0.22   0.01   0.24
```

For most standard distributions (except the exponential), there are better methods than inversion, and the happy situation exists where textbooks tend to agree about what these are. Ripley (1987, Ch. 3) is a good place to start, while the lighter version is provided by Press et al. (2007, Ch. 7). R has many of these methods built in.

References

Akaike, H. (1973). Information theory and an extension of the maximum likelihood principle. In B. Petran and F. Csaaki (Eds.), *International symposium on information theory*, Budapest: Akadeemiai Kiadi, pp. 267–281.

Berger, J. O. and L. R. Pericchi (1996). The intrinsic Bayes factor for model selection and prediction. *Journal Of The American Statistical Association 91*(433), 109–122.

Casella, G. and R. Berger (1990). *Statistical inference*. Belmont, CA: Duxbury Press.

Cox, D. R. (1992). *Planning of experiments*. New York: Wiley Classics Library.

Cox, D. R. and D. V. Hinkley (1974). *Theoretical statistics*. London: Chapman & Hall.

Davis, T. A. (2006). *Direct methods for sparse linear systems*. Philadelphia: SIAM.

Davison, A. C. (2003). *Statistical models*. Cambridge: Cambridge University Press.

De Groot, M. H. and M. J. Schervish (2002). *Probability and statistics*. Boston: Addison-Wesley.

Fahrmeir, L., T. Kneib, and S. Lang (2004). Penalized structured additive regression for space-time data: a Bayesian perspective. *Statistica Sinica 14*(3), 731–761.

Friel, N. and A. Pettitt (2008). Marginal likelihood estimation via power posteriors. *Journal of the Royal Statistical Society, Series B 70*(3), 589–607.

Gage, J. and P. Tyler (1985). Growth and recruitment of the deep-sea urchin echinus affinis. *Marine Biology 90*(1), 41–53.

Gamerman, D. and H. Lopes (2006). *Markov chain Monte Carlo: stochastic simulation for Bayesian inference*, Volume 68. Boca Raton, FL: Chapman & Hall CRC.

Gelman, A., J. B. Carlin, H. S. Stern, D. B. Dunson, A. Vehtari, and D. B. Rubin (2013). *Bayesian data analysis*. Boca Raton, FL: CRC press.

Gentle, J. (2003). *Random number generation and Monte Carlo methods* (2nd ed.). New York: Springer.

Gill, P. E., W. Murray, and M. H. Wright (1981). *Practical optimization*. London: Academic Press.

Golub, G. H. and C. F. Van Loan (2013). *Matrix computations* (4th ed.). Baltimore: Johns Hopkins University Press.

Green, P. J. (1995). Reversible jump Markov chain Monte Carlo computation and Bayesian model determination. *Biometrika 82*(4), 711–732.

Griewank, A. and A. Walther (2008). *Evaluating derivatives: principles and techniques of algorithmic differentiation*. Philadelphia: SIAM.

Grimmett, G. and D. Stirzaker (2001). *Probability and random processes* (3rd ed.). Oxford: Oxford University Press.

Gurney, W. S. C. and R. M. Nisbet (1998). *Ecological dynamics*. Oxford: Oxford University Press.

Hastie, T., R. Tibshirani, and J. Friedman (2001). *The Elements of Statistical Learning*. New York: Springer.

Kass, R. and A. Raftery (1995). Bayes factors. *Journal of the American Statistical Association 90*(430), 773–795.

Klein, J. and M. Moeschberger (2003). *Survival analysis: techniques for censored and truncated data* (2nd ed.). New York: Springer.

Marsaglia, G. (2003). Xorshift random number generators. *Journal of Statistical Software 8(14)*, 1–16.

Matsumoto, M. and T. Nishimura (1998). Mersenne twister: a 623-dimensionally equidistributed uniform pseudo-random number generator. *ACM Transactions on Modeling and Computer Simulation 8*, 3–30.

McCullagh, P. and J. A. Nelder (1989). *Generalized linear models* (2nd ed.). London: Chapman & Hall.

Neal, R. M. (2003). Slice sampling. *Annals of Statistics 31*, 705–767.

Nocedal, J. and S. Wright (2006). *Numerical optimization* (2nd ed.). New York: Springer verlag.

O'Hagan, A. (1995). Fractional Bayes factors for model comparison. *Journal of the Royal Statistical Society. Series B (Methodological) 57*(1), 99–138.

Pinheiro, J. C. and D. M. Bates (2000). *Mixed-effects models in S and S-PLUS*. New York: Springer-Verlag.

Plummer, M., N. Best, K. Cowles, and K. Vines (2006). Coda: convergence diagnosis and output analysis for MCMC. *R News 6*(1), 7–11.

Press, W., S. Teukolsky, W. Vetterling, and B. Flannery (2007). *Numerical recipes* (3rd ed.). Cambridge: Cambridge University Press.

R Core Team (2012). *R: a language and environment for statistical computing*. Vienna: R Foundation for Statistical Computing. ISBN 3-900051-07-0.

Ripley, B. D. (1987). *Stochastic simulation*. New York: Wiley.

Robert, C. (2007). *The Bayesian choice: from decision-theoretic foundations to computational implementation*. New York: Springer.

Robert, C. and G. Casella (2009). *Introducing Monte Carlo methods with R*. New York: Springer.

Roberts, G. O., A. Gelman, and W. R. Gilks (1997). Weak convergence and optimal scaling of random walk metropolis algorithms. *The Annals of Applied Probability 7*(1), 110–120.

Rue, H., S. Martino, and N. Chopin (2009). Approximate Bayesian inference for latent Gaussian models by using integrated nested Laplace approximations. *Journal of the royal statistical society: Series B 71*(2), 319–392.

Schwarz, G. (1978). Estimating the dimension of a model. *Annals of Statistics 6*(2), 461–464.

Silvey, S. D. (1970). *Statistical inference*. London: Chapman & Hall.

Spiegelhalter, D. J., N. G. Best, B. P. Carlin, and A. van der Linde (2002). Bayesian measures of model complexity and fit. *Journal of the Royal Statistical Society, Series B 64*(4), 583–639.

Steele, B. M. (1996). A modified EM algorithm for estimation in generalized mixed models. *Biometrics 52*(4), 1295–1310.

Tierney, L., R. Kass, and J. Kadane (1989). Fully exponential Laplace approximations to expectations and variances of nonpositive functions. *Journal of the American Statistical Association 84*(407), 710–716.

Watkins, D. S. (1991). *Fundamentals of matrix computation*. New York: Wiley.

Wichmann, B. and I. Hill (1982). Efficient and portable pseudo-random number generator. *Applied Statistics 31*, 188–190.

Wood, S. N. (2006). *Generalized additive models: an introduction with R*. Boca Raton, FL: CRC press.

Index